# Optimization for Communications and Networks

# Optimization for Communications and Networks

## Poompat Saengudomlert

*School of Engineering and Technology*
*Asian Institute of Technology, Thailand*

**CRC Press**
Taylor & Francis Group
an **informa** business
www.crcpress.com

6000 Broken Sound Parkway, NW
Suite 300, Boca Raton, FL 33487
270 Madison Avenue
New York, NY 10016
2 Park Square, Milton Park
Abingdon, Oxon OX14 4RN, UK

**Science Publishers**
Jersey, British Isles
Enfield, New Hampshire

Published by Science Publishers, an imprint of Edenbridge Ltd.
- St. Helier, Jersey, British Channel Islands
- P.O. Box 699, Enfield, NH 03748, USA

E-mail: *info@scipub.net*                                   Website: *www.scipub.net*

*Marketed and distributed by:*

 **CRC Press** | 6000 Broken Sound Parkway, NW Suite 300, Boca Raton, FL 33487
Taylor & Francis Group | 270 Madison Avenue New York, NY 10016
an Informa business | 2 Park Square, Milton Park Abingdon, Oxon OX14 4RN, UK
www.crcpress.com |

Copyright reserved © 2012

ISBN: 978-1-57808-724-2

**Cover Illustration:** Saroach Kuphachaka

Library of Congress Cataloging-in-Publication Data

Saengudomlert, Poompat.
   Optimization for communications and networks / Poompat Saengudomlert.
      p. cm.
   Includes bibliographical references and index.
   ISBN 978-1-57808-724-2 (hardback)
   1. Telecommunication systems—Mathematical models. 2. Mathematical optimization.
   3. Systems analysis—Mathematics.   I. Title.
   TK5102.5.S25 2011
   621.382--dc22

                                                    2011007299

To my parents
**Natee**
and
**Supreeya Saengudomlert**

# Preface

The book provides an introduction to optimization theory and its applications. It is written for senior undergraduate students and first-year graduate students of related fields. Presented example applications are for communication and network problems.

Optimization theory involves a great deal of mathematics. The book has work examples to accompany rigorous discussion so that the reader may develop intuitive understanding on relevant concepts. The materials have been developed from course notes. By attempting to cover convex, linear, and integer optimization for a one-semester course, the book focuses on fundamental concepts and techniques rather than trying to be comprehensive. In fact, the book is written with the main intention to serve as a bridge for students with no prior background in optimization to be able to access more advanced books on the subject later on.

Each main chapter starts with theoretical discussion to be followed by example applications to practical engineering problems. Exercise problems are provided at the end of each chapter, with detailed solutions given in the appendix. Thus, an interested reader can use the book as a self-study guide. Specific computer programs are also given for examples that require the use of an optimization software tool. However, since such tools are fast evolving, there is no attempt to explain them in specific details besides giving explicit examples.

I am thankful to all my teachers at Assumption College, Bangkok, Thailand; Triam Udom Suksa School, Bangkok, Thailand; Deerfield Academy, Deerfield, MA, USA; Princeton University, Princeton, NJ, USA; and Massachusetts Institute of Technology, Cambridge, MA, USA for the knowledge as well as the inspiration. I would also like to thank students who provided valuable feedback.

**Poompat Saengudomlert**
Asian Institute of Technology

# Contents

# List of Figures

# Chapter 1

# Introduction

This book discusses the fundamentals of optimization theory as well as their applications to communications and networks. The goal of the book is to provide the reader with a first look at the field of optimization, and to serve as a bridge towards more advanced textbooks on the subject. It is assumed that the reader is familiar with basic mathematical analysis and linear algebra.

When applying the theory to solve a specific problem, relevant information about a communication system or a network will be provided but without detailed discussions. In this regard, the reader with background in communications and networks may have greater appreciation of the materials. Nevertheless, the theory and the solution methods discussed in this book can be applied to a wider range of problems, e.g. operation research and economics.

Since understanding the fundamentals of optimization theory requires certain mathematical background, reviews of linear algebra and mathematical analysis are provided in the appendix. The reader is encouraged to review these mathematical tools so that he or she can easily follow the discussions in the main chapters.

The next section introduces basic components of optimization problems. The last section specifies the classes of optimization problems that are discussed in this book.

## 1.1    Components of Optimization Problems

This section introduces basic components of an optimization problem. In general, the goal of an optimization problem is to minimize or maximize some objective that is expressed as a function whose arguments are the decision variables.

Let a real function $f$ denote the *objective function*. Suppose that there are $N$ real decision variables denoted by $x_1,\ldots, x_N$. For convenience, a decision variable will be referred to simply as a *variable*. In addition, let vector $\mathbf{x} = (x_1,\ldots, x_N)$ contain all the variables. The simplest form of an optimization problem puts no restriction on the value of $\mathbf{x}$ and is referred to as *unconstrained optimization*. Mathematically, an unconstrained optimization problem has the form

$$\text{minimize } f(\mathbf{x})$$
$$\text{subject to } \mathbf{x} \in \mathbb{R}^N, \tag{1.1}$$

where $\mathbb{R}$ denotes the set of all real numbers, and $\mathbb{R}^N$ denotes the set of all $N$-dimensional real vectors.

The vector $\mathbf{x}$ in (1.1) is also called a *solution*. A solution $\mathbf{x}$ that minimizes $f(\mathbf{x})$, denoted by $\mathbf{x}^*$, is called an *optimal solution*. The value $f(\mathbf{x}^*)$ is called the *optimal cost*. Notice the usage of "an" for optimal solution and "the" for optimal cost. Note that an optimal solution may or may not exist. In addition, when an optimal solution exists, it may not be unique, i.e. multiple optimal solutions. Since maximizing $f(\mathbf{x})$ is equivalent to minimizing $-f(\mathbf{x})$, one can focus on minimization without loss of generality.

In several cases, the goal is to minimize an objective function over a set of solutions $\mathcal{F}$ that is a subset of $\mathbb{R}^N$. Such a problem is referred to as *constrained optimization* and can be expressed as

$$\text{minimize } f(\mathbf{x})$$
$$\text{subject to } \mathbf{x} \in \mathcal{F}.$$

The set $\mathcal{F}$ is called the *feasible set*. A solution $\mathbf{x}$ such that $\mathbf{x} \in \mathcal{F}$ is called a *feasible solution*. The set $\mathcal{F}$ can be continuous or discrete. An example of a continuous $\mathcal{F}$ is the set of all real numbers, i.e. $\mathbb{R}$. An example of a discrete $\mathcal{F}$ is the set of all integers denoted by $\mathbb{Z}$. Solution methods for continuous and discrete cases can be quite different in nature. The difference is significant enough that optimization over a discrete $\mathcal{F}$ is referred to as *combinatorial optimization* to be distinguished from optimization over a continuous $\mathcal{F}$.

In practical problems, one can typically express $\mathcal{F}$ using a set of expressions in terms of **x**. Each such expression is called a *constraint* for an optimization problem. For example, if the goal is to minimize $f(x)$ over the set of nonnegative real numbers denoted by $\mathbb{R}^+$, then $\mathcal{F}$ can be described using a single constraint $x \geq 0$, i.e.

$$\text{minimize } f(x)$$
$$\text{subject to } x \geq 0.$$

Throughout this book, it is assumed that $\mathcal{F}$ can be described using a combination (not necessarily all) of inequality constraints, equality constraints, and integer constraints

$$\forall l \in \{1,..., L\}, \ g_l(\mathbf{x}) \leq 0,$$
$$\forall m \in \{1,..., M\}, \ h_m(\mathbf{x}) = 0,$$
$$\mathbf{x} \in \mathbb{Z}^N, \tag{1.2}$$

where $L$ and $M$ are the numbers of inequality and equality constraints, $g_l$ and $h_m$ are real constraint functions, and $\mathbb{Z}^N$ denotes the set of all $N$-dimensional integer vectors. Given the objective and constraint functions, it is implicitly assumed that only solutions in the domain set of all functions $f$, $g_l$, and $h_m$ are considered. The above assumption on $\mathcal{F}$ is a mild one since in a wide range of practical problems $\mathcal{F}$ can be expressed as described above.

## 1.2  Classes of Optimization Problems

This book discusses three classes of optimization problems.

1. *Convex optimization*: In a convex optimization problem, the objective function $f$ is a convex function. The feasible set $\mathcal{F}$ is a convex set. (The definitions of a convex set and a convex function are presented in chapter 2.) There is no integer constraint on a solution **x**.

2. *Linear optimization*: In a linear optimization problem, the objective and constraint functions $f$, $g_l$, $h_m$ are linear. (The definition of a linear function is presented in chapter 3.) There is no integer constraint on a solution **x**.

3. *Integer linear optimization*: In an integer linear optimization problem, the objective and constraint functions $f$, $g_l$, $h_m$ are linear. In addition, there is a constraint that **x** must be an integer vector. Note that the only difference between this case and linear optimization is the integer constraint.

Since linear optimization is a special case of convex optimization, convex optimization is discussed first. The results obtained for convex optimization can then be applied to linear optimization.

A large number of integer linear optimization problems are known to be computationally difficult to solve, and remain open problems for research. For this reason, heuristic solutions are developed to solve these problems in practice. Solution methods for convex optimization and linear optimization are often useful in constructing such heuristics.

A more general class of optimization problems involve nonlinear optimization, which does not belong to any of the above three classes. Nevertheless, solution methods for nonlinear optimization are often based on those for convex optimization. Hence, the discussions in this book will serve as a useful basis for the reader who study nonlinear optimization later on.

It is worth noting that the term "optimization" is often used interchangeably with the term "programing". For example, it is common to read about a "linear programing (LP)" problem instead of a "linear optimization" problem. The two terms are equivalent. Since optimization problems cannot in general be solved to obtain closed form solutions, computer programing is done to numerically compute optimal solutions. Hence, the term "programing" is used to reflect how such problems are solved.

Finally, this book is written to be as instructive as possible. Therefore, most results are justified by proofs. However, the proofs that are marked with '†' can be skipped by readers who may not need the justifications.

# Chapter 2

# Convex Optimization

This chapter focuses on convex optimization problems. These optimization problems have an important property that a local optimal solution is also a global optimal solution. The first few sections provide theoretical development of dual problems, Lagrange multipliers, and Karush-Kuhn-Tucker (KKT) conditions. These theoretical discussions provide analytical tools that can be used to obtain closed form solutions for some problems. For other problems that do not have closed form solutions, commonly used numerical algorithms are discussed. Finally, specific applications of convex optimization are given, including power allocation in multicarrier communications and routing in packet switching networks.

## 2.1   Convex Sets and Convex Functions

**Convex Sets**

Let $N$ be a positive integer. Let $\mathbf{x}, \mathbf{y} \in \mathbb{R}^N$ with $\mathbf{x} \neq \mathbf{y}$. A point $\mathbf{z} \in \mathbb{R}^N$ is a *convex combination* of $\mathbf{x}$ and $\mathbf{y}$ if $\mathbf{z}$ can be expressed as

$$\mathbf{z} = \alpha\mathbf{x} + (1 - \alpha)\mathbf{y}, \ \alpha \in [0, 1]. \qquad (2.1)$$

Geometrically, if $\mathbf{z}$ is a convex combination of $\mathbf{x}$ and $\mathbf{y}$, then $\mathbf{z}$ lies somewhere on a line segment connecting $\mathbf{x}$ and $\mathbf{y}$, as illustrated in figure 2.1.

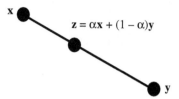

**Figure 2.1**    Convex combination of **x** and **y**.

A set $\mathcal{X} \subset \mathbb{R}^N$ is a *convex set* if, for any two points **x**, **y** $\in$ $\mathcal{X}$, any convex combination of **x** and **y** is also in $\mathcal{X}$. Geometrically, for any line segment connecting two points in the set, all the points on this line segment are contained in the set. Figure 2.2 shows examples of convex and non-convex sets in $\mathbb{R}^2$.

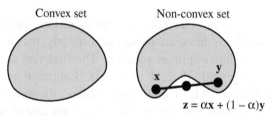

**Figure 2.2**    Convex and non-convex sets in $\mathbb{R}^2$

## Convex Functions

Let $\mathcal{X}$ be a convex set in $\mathbb{R}^N$. Consider a real function $f$ defined on $\mathcal{X}$. Such a function $f$ is *convex* if, for all **x**, **y** $\in$ $\mathcal{X}$ and $\alpha \in (0, 1)$,

$$f(\alpha \mathbf{x} + (1 - \alpha)\mathbf{y}) \le \alpha f(\mathbf{x}) + (1 - \alpha) f(\mathbf{y}). \tag{2.2}$$

If the above condition is satisfied with strict inequality, then $f$ is *strictly convex*. In addition, $f$ is *strictly concave* if $-f$ is strictly convex. Note that a convex function is always defined on a convex set. The following theorems state basic properties of convex functions.

**Theorem 2.1 (First-order condition)**    Let $\mathcal{X} \subset \mathbb{R}^N$ be a convex set, and $f$ be a real and differentiable function defined on $\mathcal{X}$. Then, $f$ is convex if and only if

$$f(\mathbf{y}) \ge f(\mathbf{x}) + \nabla f(\mathbf{x})^{\mathrm{T}} (\mathbf{y} - \mathbf{x}) \text{ for all } \mathbf{x}, \mathbf{y} \in \mathcal{X}.$$

**†Proof** Suppose that $f$ is convex. Interchanging the roles of **x** and **y** in (2.2) yields $f(\alpha\mathbf{y} + (1 - \alpha)\mathbf{x}) \leq \alpha f(\mathbf{y}) + (1 - \alpha) f(\mathbf{x})$ for all $\mathbf{x}, \mathbf{y} \in \mathcal{X}$ and $\alpha \in (0, 1)$. The inequality can be written as

$$f(\mathbf{y}) - f(\mathbf{x}) \geq \frac{f(\mathbf{x} + \alpha(\mathbf{y} - \mathbf{x})) - f(\mathbf{x})}{\alpha}$$

Taking the limit as $\alpha \to 0$ yields the desired inequality, i.e.

$$f(\mathbf{y}) - f(\mathbf{x}) \geq \lim_{\alpha \to 0} \frac{f(\mathbf{x} + \alpha(\mathbf{y} - \mathbf{x})) - f(\mathbf{x})}{\alpha} = \nabla f(\mathbf{x})^{\mathrm{T}}(\mathbf{y} - \mathbf{x}).$$

For the converse part, suppose the inequality in the theorem statement holds. For arbitrary $\mathbf{x}, \mathbf{y} \in \mathcal{X}$ and $\alpha \in (0, 1)$, let $\mathbf{z} = \alpha\mathbf{x} + (1 - \alpha)\mathbf{y}$. Using the inequality twice yields

$$f(\mathbf{x}) \geq f(\mathbf{z}) + \nabla f(\mathbf{z})^{\mathrm{T}}(\mathbf{x} - \mathbf{z})$$
$$f(\mathbf{y}) \geq f(\mathbf{z}) + \nabla f(\mathbf{z})^{\mathrm{T}}(\mathbf{y} - \mathbf{z})$$

Multiplying the first and second inequalities by $\alpha$ and $1 - \alpha$ together with adding the results yields

$$\alpha f(\mathbf{x}) + (1 - \alpha) f(\mathbf{y}) \geq f(\mathbf{z}) + \nabla f(\mathbf{z})^{\mathrm{T}} \underbrace{(\alpha\mathbf{x} + (1 - \alpha)\mathbf{y} - \mathbf{z})}_{= 0}$$

$$= f(\mathbf{z}) = f(\alpha\mathbf{x} + (1 - \alpha)\mathbf{y}),$$

which implies that $f$ is convex. □

Figure 2.3 illustrates the statement of theorem 2.1 for $\mathcal{X} = \mathbb{R}$. Since the expression $f(\mathbf{x}) + \nabla f(\mathbf{x})^{\mathrm{T}}(\mathbf{y} - \mathbf{x})$ is the first-order Taylor series approximation of $f(\mathbf{y})$ (see (A.6)), theorem 2.1 indicates that the first-order Taylor series approximation always underestimates a convex function.

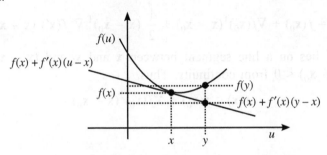

**Figure 2.3** Illustration of theorem 2.1 for $\mathcal{X} = \mathbb{R}$.

**Theorem 2.2 (Second-order condition)**   Let $\mathcal{X} \subset \mathbb{R}^N$ be a convex set, and $f$ be a real and twice differentiable function defined on $\mathcal{X}$. Then, $f$ is convex if and only if its Hessian $\nabla^2 f$ is positive semidefinite in $\mathcal{X}$, i.e.

$$\nabla^2 f(\mathbf{x}) \geq 0 \quad \text{for all } \mathbf{x} \in \mathcal{X}.$$

†**Proof**   Suppose that $\nabla^2 f(\mathbf{x}) \geq 0$ for all $\mathbf{x} \in \mathcal{X}$. Consider the second-order Tayor series expansion of $f(\mathbf{x})$ with respect to $\mathbf{x}_0 \in \mathcal{X}$

$$f(\mathbf{x}) = f(\mathbf{x}_0) + \nabla f(\mathbf{x}_0)^{\mathrm{T}}(\mathbf{x} - \mathbf{x}_0) + \frac{1}{2}(\mathbf{x} - \mathbf{x}_0)^{\mathrm{T}} \nabla^2 f(\mathbf{x}')(\mathbf{x} - \mathbf{x}_0),$$

where $\mathbf{x}'$ lies on a line segment between $\mathbf{x}$ and $\mathbf{x}_0$. Since $\nabla^2 f(\mathbf{x}') \geq 0$,

$$f(\mathbf{x}) \geq f(\mathbf{x}_0) + \nabla f(\mathbf{x}_0)^{\mathrm{T}}(\mathbf{x} - \mathbf{x}_0). \tag{2.3}$$

Let $\alpha \in (0, 1)$ and $\mathbf{x}_1, \mathbf{x}_2 \in \mathcal{X}$. By setting $\mathbf{x}_0 = \alpha \mathbf{x}_1 + (1 - \alpha)\mathbf{x}_2$ and $\mathbf{x} = \mathbf{x}_1$, (2.3) can be written as

$$f(\mathbf{x}_1) \geq f(\alpha \mathbf{x}_1 + (1 - \alpha)\mathbf{x}_2) + \nabla f(\mathbf{x}_0)^{\mathrm{T}}((1 - \alpha)(\mathbf{x}_1 - \mathbf{x}_2)). \tag{2.4}$$

By setting $\mathbf{x}_0 = \alpha \mathbf{x}_1 + (1 - \alpha)\mathbf{x}_2$ and $\mathbf{x} = \mathbf{x}_2$, (2.3) can be written as

$$f(\mathbf{x}_2) \geq f(\alpha \mathbf{x}_1 + (1 - \alpha)\mathbf{x}_2) - \nabla f(\mathbf{x}_0)^{\mathrm{T}}(\alpha(\mathbf{x}_1 - \mathbf{x}_2)). \tag{2.5}$$

Adding (2.4) multiplied by $\alpha$ and (2.5) multiplied by $1 - \alpha$ yields

$$f(\alpha \mathbf{x}_1 + (1 - \alpha)\mathbf{x}_2) \leq \alpha f(\mathbf{x}_1) + (1 - \alpha) f(\mathbf{x}_2),$$

which implies that $f(\mathbf{x})$ is convex.

The proof for the converse part can be written based on contradiction. Suppose that there is $\mathbf{x}_0 \in \mathcal{X}$ such that $\nabla^2 f(\mathbf{x}_0)$ is not positive semidefinite. It follows that, for some $\mathbf{x}$ close to $\mathbf{x}_0$, $(\mathbf{x} - \mathbf{x}_0)^{\mathrm{T}} \nabla^2 f(\mathbf{x}_0)(\mathbf{x} - \mathbf{x}_0) < 0$. Consider the second-order Taylor series expansion

$$f(\mathbf{x}) = f(\mathbf{x}_0) + \nabla f(\mathbf{x}_0)^{\mathrm{T}}(\mathbf{x} - \mathbf{x}_0) + \frac{1}{2}(\mathbf{x} - \mathbf{x}_0)^{\mathrm{T}} \nabla^2 f(\mathbf{x}')(\mathbf{x} - \mathbf{x}_0),$$

where $\mathbf{x}'$ lies on a line segment between $\mathbf{x}$ and $\mathbf{x}_0$ and $(\mathbf{x} - \mathbf{x}_0)^{\mathrm{T}} \nabla^2 f(\mathbf{x}')(\mathbf{x} - \mathbf{x}_0) < 0$ from continuity. Thus,

$$f(\mathbf{x}) < f(\mathbf{x}_0) + \nabla f(\mathbf{x}_0)^{\mathrm{T}}(\mathbf{x} - \mathbf{x}_0) \tag{2.6}$$

Now consider a point $\mathbf{x}_0 + \alpha(\mathbf{x} - \mathbf{x}_0)$ for $\alpha \in (0, 1)$. Using the fact that

$$\lim_{\alpha \to 0} \frac{f(\mathbf{x}_0 + \alpha(\mathbf{x} - \mathbf{x}_0)) - f(\mathbf{x}_0)}{\alpha} = \nabla f(\mathbf{x}_0)^{\mathrm{T}}(\mathbf{x} - \mathbf{x}_0),$$

(2.6) can be written as

$$f(\mathbf{x}) < f(\mathbf{x}_0) + \lim_{\alpha \to 0} \frac{f(\mathbf{x}_0 + \alpha(\mathbf{x} - \mathbf{x}_0)) - f(\mathbf{x}_0)}{\alpha}$$

$$= \lim_{\alpha \to 0} \frac{-(1 - \alpha) f(\mathbf{x}_0) + f(\alpha \mathbf{x} + (1 - \alpha) \mathbf{x}_0)}{\alpha}$$

which implies that there exists some $\alpha > 0$ such that $\alpha f(\mathbf{x}) + (1 - \alpha) f(\mathbf{x}_0)$ $< f(\alpha \mathbf{x} + (1 - \alpha) \mathbf{x}_0)$, contradicting the assumption that $f$ is convex. □

## 2.2   Properties of Convex Optimization

In general, a *convex optimization problem* can be expressed in the following form.

$$\begin{array}{l} \text{minimize } f(\mathbf{x}) \\ \text{subject to } \forall l \in \{1, ..., L\}, g_l(\mathbf{x}) \leq 0 \\ \qquad\qquad \forall m \in \{1, ..., M\}, h_m(\mathbf{x}) = 0 \end{array} \qquad (2.7)$$

It is assumed that $\mathbf{x} \in \mathbb{R}^N$, and $f, g_l, h_m$ are real functions defined on $\mathbb{R}^N$. In addition, the feasible set is assumed to be a convex set. The objective function $f$ is assumed to be a convex function on the set

$$\mathcal{X} = \mathcal{D}(f) \cap \left( \bigcap_{l=1}^{L} \mathcal{D}(g_l) \right) \cap \left( \bigcap_{m=1}^{M} \mathcal{D}(h_m) \right), \qquad (2.8)$$

where $\mathcal{D}(f)$ denotes the domain set of function $f$.

A feasible solution is a point $\mathbf{x} \in \mathcal{X}$ that satisfies all the constraints. Let $\mathcal{F}$ denote the feasible set, i.e. the set of all feasible solutions. A point $\mathbf{x}^* \in \mathcal{F}$ is a *local minimum* if there exists an $\varepsilon > 0$ such that

$$f(\mathbf{x}^*) \leq f(\mathbf{x}) \quad \text{for all } \mathbf{x} \in \mathcal{F} \text{ with } \|\mathbf{x}^* - \mathbf{x}\| < \varepsilon.$$

Roughly speaking, a local minimum has a cost no greater than those of its feasible neighbors. A point $\mathbf{x}^*$ is a *global minimum* if

$$f(\mathbf{x}^*) \leq f(\mathbf{x}) \quad \text{for all } \mathbf{x} \in \mathcal{F}.$$

The following theorem states one key property of convex optimization.

**Theorem 2.3 (Equivalence of local and global minima)**   For the convex optimization problem in (2.7), a local minimum is also a global minimum. In addition, if $f$ is strictly convex, the global minimum is unique.

**Proof**   The first statement can be proved by contradiction. Suppose that a local minimum $\mathbf{x}$ is not a global minimum. Then, there is a point $\mathbf{y} \in \mathcal{F}$ with $\mathbf{y} \neq \mathbf{x}$ such that $f(\mathbf{y}) < f(\mathbf{x})$. However, since $f$ is convex, it follows that

$$f(\alpha \mathbf{x} + (1 - \alpha)\mathbf{y}) \leq \alpha f(\mathbf{x}) + (1 - \alpha) f(\mathbf{y}) < f(\mathbf{x})$$

for any $\alpha \in (0, 1)$, contradicting the assumption that $\mathbf{x}$ is a local minimum.

The second statement can also be proved by contradiction. Suppose that there are two global minima $\mathbf{x}$ and $\mathbf{y}$. Consider a feasible solution $\mathbf{z} = (\mathbf{x} + \mathbf{y})/2$. By strict convexity, $f(\mathbf{z}) < (f(\mathbf{x}) + f(\mathbf{y}))/2 = f(\mathbf{x})$, contradicting the assumption that $\mathbf{x}$ is a global minimum.   □

For a differentiable convex function $f$, the next theorem states a necessary and sufficient condition for $\mathbf{x} \in \mathcal{F}$ to be a global minimum.
**Theorem 2.4 (Optimality condition)**   Suppose that, for the convex optimization problem in (2.7), $f$ is differentiable. Then $\mathbf{x}^* \in \mathcal{F}$ is a global minimum if and only if

$$\nabla f(\mathbf{x}^*)^{\mathrm{T}}(\mathbf{x} - \mathbf{x}^*) \geq 0 \quad \text{for all } \mathbf{x} \in \mathcal{F}.$$

†**Proof**   From theorem 2.1, $f(\mathbf{x}) \geq f(\mathbf{x}^*) + \nabla f(\mathbf{x}^*)^{\mathrm{T}}(\mathbf{x} - \mathbf{x}^*)$ for all $\mathbf{x} \in \mathcal{F}$. Suppose that the theorem statement holds. It follows that

$$f(\mathbf{x}) \geq f(\mathbf{x}^*) + \nabla f(\mathbf{x}^*)^{\mathrm{T}}(\mathbf{x} - \mathbf{x}^*) \geq f(\mathbf{x}^*) \quad \text{for all } \mathbf{x} \in \mathcal{F},$$

which implies that $\mathbf{x}^*$ is a global minimum.

The converse part is now proved by contradiction. Suppose that $\mathbf{x}^*$ is a global minimum, and there is an $\mathbf{x} \in \mathcal{F}$ such that $\nabla f(\mathbf{x}^*)(\mathbf{x} - \mathbf{x}^*) < 0$. Let $\mathbf{z} = \alpha \mathbf{x} + (1 - \alpha)\mathbf{x}^*$, where $\alpha \in (0, 1)$. From the convexity of $\mathcal{F}$, $\mathbf{z} \in \mathcal{F}$. From the mean value theorem (see theorem A.4),

$$f(\mathbf{z}) = f(\mathbf{x}^* + \alpha(\mathbf{x} - \mathbf{x}^*)) = f(\mathbf{x}^*) + \alpha\nabla f(\mathbf{x}^* + \beta\alpha(\mathbf{x} - \mathbf{x}^*))^T(\mathbf{x} - \mathbf{x}^*)$$

for some $\beta \in [0, 1]$. For a sufficiently small $\alpha > 0$, $\nabla f(\mathbf{x}^* + \beta\alpha\,(\mathbf{x} - \mathbf{x}^*))^T(\mathbf{x} - \mathbf{x}^*) < 0$ by continuity. It follows that $f(\mathbf{z}) < f(\mathbf{x}^*)$ for this value of $\alpha$, contradicting the assumption that $\mathbf{x}^*$ is a global minimum.    □

Figure 2.4 illustrates the statement of theorem 2.4 by indicating that the gradient $\nabla f(\mathbf{x}^*)$ and the feasible variation $\mathbf{x} - \mathbf{x}^*$, $\mathbf{x} \in \mathcal{F}$, make an angle of no more than 90 degree. In addition, note that figure 2.4 uses the *contour lines* to illustrate the function values for $\mathcal{F} \subset \mathbb{R}^2$; a countour line refers to a set of points that yield the same objective function value.

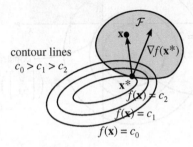

**Figure 2.4**   Illustration of theorem 2.4 for $\mathcal{F} \subset \mathbb{R}^2$.

**Example 2.1:**   Consider minimizing $f(x) = (x + 1)^2$ subject to $x \geq 0$. Since $df(x)/dx = 2(x + 1)$ is positive for all $x \geq 0$, the only solution that satisfies the condition in theorem 2.4 is $x^* = 0$, as shown in figure 2.5. More specifically, if $x > 0$, then we can take $x/2$ and $x$ to obtain the feasible variation such that $df(x)/dx \times (x/2 - x) < 0$. Hence, $x^* = 0$ is the unique optimal solution with $f(0) = 1$ as the optimal cost.    □

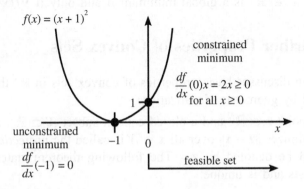

**Figure 2.5**   Example application of theorem 2.4 for $\mathcal{F} \subset \mathbb{R}$

**Example 2.2:** Let $\mathbf{x} \in \mathbb{R}^2$. Consider minimizing $f(\mathbf{x}) = (x_1 + 1)^2 + (x_2 + 1)^2$ subject to $\mathbf{x} \geq \mathbf{0}$. Since $\nabla f(\mathbf{x}) = \begin{bmatrix} 2(x_1 + 1) \\ 2(x_2 + 1) \end{bmatrix} \geq \mathbf{0}$ for all $\mathbf{x} \geq \mathbf{0}$, the only solution that satisfies the condition in theorem 2.4 is $\mathbf{x}^* = \mathbf{0}$, as shown in figure 2.6. Hence, $\mathbf{x}^* = \mathbf{0}$ is the unique optimal solution with $f(\mathbf{0}) = 2$ as the optimal cost. □

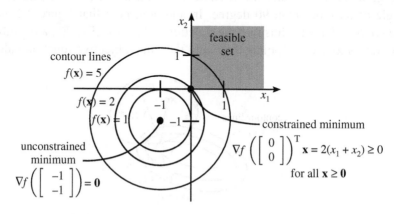

**Figure 2.6**   Example application of theorem 2.4 for $\mathcal{F} \subset \mathbb{R}^2$.

For unconstrained convex optimization in $\mathbb{R}^N$, since the feasible set is $\mathbb{R}^N$, the only way the condition in theorem 2.4 is satisfied is to have $\nabla f(\mathbf{x}^*) = \mathbf{0}$. This condition is formally stated below.

**Corollary 2.1 (Optimality condition for unconstrained optimization)** For unconstrained convex optimization with differentiable objective function $f$, $\mathbf{x}^* \in \mathbb{R}^N$ is a global minimum if and only if $\nabla f(\mathbf{x}^*) = \mathbf{0}$.

## 2.3   Further Properties of Convex Sets

This section discusses basic properties of convex sets in $\mathbb{R}^N$ that can be understood by geometric visualization.

For a point $\mathbf{z} \in \mathbb{R}^N$ and a closed convex subset $\mathcal{X} \subset \mathbb{R}^N$, a point in $\mathcal{X}$ that minimizes $\|\mathbf{z} - \mathbf{x}\|$ over all $\mathbf{x} \in \mathcal{X}$ is called the *projection of* $\mathbf{z}$ *on* $\mathcal{X}$ and will be denoted by $\mathbf{z}_{|\mathcal{X}}$. The following theorem states that $\mathbf{z}_{|\mathcal{X}}$ always exists and is unique.

**Theorem 2.5 (Projection theorem)**

1. For any given $\mathbf{z} \in \mathbb{R}^N$ and a closed convex subset $\mathcal{X} \subset \mathbb{R}^N$, there is a unique vector $\mathbf{z}_{|\mathcal{X}} \in \mathcal{X}$ that minimizes $\|\mathbf{z} - \mathbf{x}\|$ over all $\mathbf{x} \in \mathcal{X}$.

2. A vector $\mathbf{x}'$ is equal to the projection vector $\mathbf{z}_{|\mathcal{X}}$ if and only if

$$(\mathbf{z} - \mathbf{x}')^{\mathrm{T}}(\mathbf{x} - \mathbf{x}') \leq 0 \quad \text{for all } \mathbf{x} \in \mathcal{X}. \tag{2.9}$$

The proof of theorem 2.5 relies on the following lemma.

**Lemma 2.1 (Convexity of intersections and closures of convex sets)**

1. Let $\{\mathcal{A}_i | i \in \mathcal{I}\}$ be a collection of convex sets, where $\mathcal{I}$ denotes the set of indices. The set $\mathcal{X} = \cap_{i \in \mathcal{I}} \mathcal{A}_i$ is convex.

2. Let $\bar{\mathcal{X}}$ denote the closure of $\mathcal{X}$. If $\mathcal{X}$ is a convex set, then so is $\bar{\mathcal{X}}$.

†**Proof**

1. Let $\mathbf{x}, \mathbf{y} \in \mathcal{X}$. Let $\mathbf{z} = \alpha \mathbf{x} + (1 - \alpha)\mathbf{y}$, where $\alpha \in [0, 1]$. For any $i \in \mathcal{I}$, since $\mathbf{x}, \mathbf{y} \in \mathcal{A}_i$, it follows that $\mathbf{z} \in \mathcal{A}_i$ by the convexity of $\mathcal{A}_i$. Hence, $\mathbf{z} \in \mathcal{X}$.

2. Since $\bar{\mathcal{X}}$ can be expressed as $\bar{\mathcal{X}} = \cap_{\varepsilon > 0} \{\mathbf{x} + \mathbf{y} | \mathbf{x} \in X, \|\mathbf{y}\| < \varepsilon\}$, statement 1 implies that $\bar{\mathcal{X}}$ is a convex set. □

†**Proof of theorem 2.5**

1. Consider an arbitrary $\mathbf{y} \in \mathcal{X}$. Minimizing $\|\mathbf{z} - \mathbf{x}\|$ subject to $\mathbf{x} \in \mathcal{X}$ is equivalent to minimizing $\|\mathbf{z} - \mathbf{x}\|$ subject to $\mathbf{x} \in \mathcal{X}'$, where

$$\mathcal{X}' = \mathcal{X} \cap \{\mathbf{x} \in \mathbb{R}^N | \|\mathbf{z} - \mathbf{x}\| \leq \|\mathbf{y} - \mathbf{x}\|\}.$$

This new problem is in turn equivalent to minimizing $\|\mathbf{z} - \mathbf{x}\|^2$ subject to $\mathbf{x} \in \mathcal{X}$. Since $\mathcal{X}'$ is closed and bounded, it is compact. Since $f(\mathbf{x}) = \|\mathbf{z} - \mathbf{x}\|^2$ is a continuous function, the Weierstrass theorem (see theorem A.3) implies that there is a minimum point.

It is straightforward to verify that $f(\mathbf{x}) = \|\mathbf{z} - \mathbf{x}\|^2$ is strictly convex. Theorem 2.3 then specifies that the minimum point is unique.

2. The statement results from applying statement 1 and theorem 2.4 to the problem of minimizing $\|\mathbf{z} - \mathbf{x}\|^2$ subject to $\mathbf{x} \in \mathcal{X}$. □

Figure 2.7 illustrates statement 2 of theorem 2.5, which indicates that the projection error vector $\mathbf{z} - \mathbf{z}_{|\mathcal{X}}$ and vector $\mathbf{x} - \mathbf{z}_{|\mathcal{X}}$ for any $\mathbf{x} \in \mathcal{X}$ make an angle of at least 90 degree.

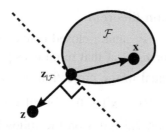

**Figure 2.7**    Illustration of theorem 2.5 for $\mathcal{F} \subset \mathbb{R}^2$.

The subset of $\mathbb{R}^N$ of the form $\{\mathbf{x} \in \mathbb{R}^N | \mathbf{a}^T\mathbf{x} = c\}$ is called the *hyperplane* defined by $\mathbf{a}$ and $c$. A hyperplane separates $\mathbb{R}^N$ into two disjoint subsets $\{\mathbf{x} \in \mathbb{R}^N | \mathbf{a}^T\mathbf{x} < c\}$ and $\{\mathbf{x} \in \mathbb{R}^N | \mathbf{a}^T\mathbf{x} \geq c\}$. Each point in the set $\{\mathbf{x} \in \mathbb{R}^N | \mathbf{a}^T\mathbf{x} \geq c\}$ is said to be *supported* by the hyperplane defined by $\mathbf{a}$ and $c$. The following two theorems specify basic properties of hyperplanes with respect to convex sets.

**Theorem 2.6 (Supporting hyperplane theorem)**    Let $\mathcal{X} \subset \mathbb{R}^N$ be a convex set and $\mathbf{x}'$ be a point that is not an interior point of $\mathcal{X}$. Then, there exists a hyperplane containing $\mathbf{x}'$ that supports $\mathcal{X}$. More specifically, there exists a nonzero vector $\mathbf{a} \in \mathbb{R}^N$ such that

$$\mathbf{a}^T\mathbf{x} \geq \mathbf{a}^T\mathbf{x}' \quad \text{for all } \mathbf{x} \in \mathcal{X}.$$

†**Proof**    Let $\bar{\mathcal{X}}$ be the closure of $\mathcal{X}$. Being a closure of a convex set, $\bar{\mathcal{X}}$ is closed and convex.

Since $\mathbf{x}'$ is not an interior point of $\mathcal{X}$, there exists a sequence $\mathbf{x}^1$, $\mathbf{x}^2$, ... outside $\bar{\mathcal{X}}$ that converges to $\mathbf{x}'$. Let $\bar{\mathbf{x}}^1$, $\bar{\mathbf{x}}^2$, ... be the projections of $\mathbf{x}^1$, $\mathbf{x}^2$, ... on $\bar{\mathcal{X}}$. From theorem 2.5,

$$(\mathbf{x}^k - \bar{\mathbf{x}}^k)^T (\mathbf{x} - \bar{\mathbf{x}}^k) \leq 0 \quad \text{for all } \mathbf{x} \in \bar{\mathcal{X}}.$$

It follows that

$$(\bar{\mathbf{x}}^k - \mathbf{x}^k)^T \mathbf{x} \geq (\bar{\mathbf{x}}^k - \mathbf{x}^k)^T \bar{\mathbf{x}}^k = \underbrace{(\bar{\mathbf{x}}^k - \mathbf{x}^k)^T (\bar{\mathbf{x}}^k - \mathbf{x}^k)}_{= \|\bar{\mathbf{x}}^k - \mathbf{x}^k\|^2 \geq 0}$$

$$+ (\bar{\mathbf{x}}^k - \mathbf{x}^k)^T \mathbf{x}^k \geq (\bar{\mathbf{x}}^k - \mathbf{x}^k)^T \mathbf{x}^k$$

or equivalently

$$\mathbf{a}^{kT} \mathbf{x} \geq \mathbf{a}^{kT} \mathbf{x}^k$$

for all $k \in \{1, 2, \ldots\}$ and $\mathbf{x} \in \bar{\mathcal{X}}$, where $\mathbf{a}^k = \dfrac{\bar{\mathbf{x}}^k - \mathbf{x}^k}{\|\bar{\mathbf{x}}^k - \mathbf{x}^k\|}$. Since $\|\mathbf{a}^k\| = 1$ for all $k$, as $\mathbf{x}^k$ converges to $\mathbf{x}'$, there is a subsequence of $\mathbf{a}^k$ that converges to a nonzero limit $\mathbf{a}$. Taking the limit of $\mathbf{a}^{k\mathrm{T}} \mathbf{x} \geq \mathbf{a}^{k\mathrm{T}} \mathbf{x}^k$ as $k \to \infty$ yields the theorem statement. □

Figure 2.8 illustrates the supporting hyperplane theorem for $\mathbb{R}^2$.

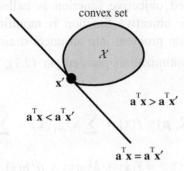

**Figure 2.8** Illustration of the supporting hyperplane theorem for $\mathbb{R}^2$

**Theorem 2.7 (Separating hyperplane theorem)** Let $\mathcal{X}_1, \mathcal{X}_2 \subset \mathbb{R}^N$ be two disjoint convex sets. Then there exists a hyperplane that separates them. More specifically, there is a nonzero vector $\mathbf{a} \in \mathbb{R}^N$ such that

$$\mathbf{a}^{\mathrm{T}}\mathbf{x}_1 \geq \mathbf{a}^{\mathrm{T}}\mathbf{x}_2 \quad \text{for all} \quad \mathbf{x}_1 \in \mathcal{X}_1,\ \mathbf{x}_2 \in \mathcal{X}_2.$$

<sup>†</sup>**Proof** Consider the set $\mathcal{A} = \{\mathbf{x}_1 - \mathbf{x}_2 | \mathbf{x}_1 \in \mathcal{X}_1,\ \mathbf{x}_2 \in \mathcal{X}_2\}$. It is straightforward to verify that $\mathcal{A}$ is a convex set. Since $\mathcal{X}_1$ and $\mathcal{X}_2$ are disjoint, the zero vector does not belong to $\mathcal{A}$. From the supporting hyperplane theorem, there exists a nonzero vector $\mathbf{a} \in \mathbb{R}^N$ such that

$$\mathbf{a}^{\mathrm{T}}\mathbf{x} \geq 0 \quad \text{for all} \quad \mathbf{x} \in \mathcal{A},$$

which is equivalent to the theorem statement. □

Figure 2.9 illustrates the separating hyperplane theorem for $\mathbb{R}^2$.

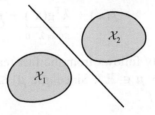

**Figure 2.9** Illustration of the separating hyperplane theorem for $\mathbb{R}^2$.

## 2.4   Dual Problems

Consider the convex optimization problem in (2.7), which will be referred to as the *primal problem*. For convenience, define $\mathbf{g}(\mathbf{x}) = (g_1(\mathbf{x}), ..., g_L(\mathbf{x}))$ and $\mathbf{h}(\mathbf{x}) = (h_1(\mathbf{x}), ..., h_M(\mathbf{x}))$. The *Lagrange multiplier method* is one approach of finding an optimal solution. In this method, the constraints $\mathbf{g}(\mathbf{x}) \leq 0$ and $\mathbf{h}(\mathbf{x}) = 0$ are incorporated into the objective function. The modified objective function is called the *Lagrangian*. Roughly speaking, the objective function is modified to turn a given constrained optimization problem into an unconstrained one.

For the convex optimization problem in (2.7), the Lagrangian is defined as

$$\Lambda(\mathbf{x}, \boldsymbol{\lambda}, \boldsymbol{\mu}) = f(\mathbf{x}) + \sum_{l=1}^{L} \lambda_l \, g_l(\mathbf{x}) + \sum_{m=1}^{M} \mu_m h_m(\mathbf{x})$$

$$= f(\mathbf{x}) + \boldsymbol{\lambda}^{\mathrm{T}} \mathbf{g}(\mathbf{x}) + \boldsymbol{\mu}^{\mathrm{T}} \mathbf{h}(\mathbf{x}), \tag{2.10}$$

where $\boldsymbol{\lambda} = (\lambda_1, ..., \lambda_L)$ and $\boldsymbol{\mu} = (\mu_1, ..., \mu_M)$ are called *dual variables*. Define the *dual function*, denoted by $q$, as

$$q(\boldsymbol{\lambda}, \boldsymbol{\mu}) = \inf_{\mathbf{x} \in \mathcal{X}} \Lambda(\mathbf{x}, \boldsymbol{\lambda}, \boldsymbol{\mu}), \tag{2.11}$$

where $\mathcal{X}$ is the domain set defined in (2.8). Note that dual variables are the arguments of the dual function.

Let $\mathbf{x}^*$ and $f^*$ denote an optimal solution and the optimal cost of the primal problem. For any $\boldsymbol{\lambda} \geq 0$ and $\boldsymbol{\mu} \in \mathbb{R}^M$, the dual function in (2.11) yields a lower bound on $f^*$. To see this, consider an arbitrary primal feasible solution $\mathbf{x}' \in \mathcal{F}$. Using $\mathbf{x}'$, the dual function can be bounded by

$$q(\boldsymbol{\lambda}, \boldsymbol{\mu}) = \inf_{\mathbf{x} \in \mathcal{X}} \Lambda(\mathbf{x}, \boldsymbol{\lambda}, \boldsymbol{\mu}) \leq \Lambda(\mathbf{x}', \boldsymbol{\lambda}, \boldsymbol{\mu})$$

$$= f(\mathbf{x}') + \underbrace{\boldsymbol{\lambda}^{\mathrm{T}} \mathbf{g}(\mathbf{x}')}_{\leq 0 \text{ for } \boldsymbol{\lambda} \geq 0} + \underbrace{\boldsymbol{\mu}^{\mathrm{T}} \mathbf{h}(\mathbf{x}')}_{= 0} \leq f(\mathbf{x}'),$$

where the last inequality follows from the fact that $\boldsymbol{\lambda}^{\mathrm{T}} \mathbf{g}(\mathbf{x}') \leq 0$ for $\boldsymbol{\lambda} \geq 0$ and $\boldsymbol{\mu}^{\mathrm{T}} \mathbf{h}(\mathbf{x}') = 0$ for any $\boldsymbol{\mu} \in \mathbb{R}^M$. Since $q(\boldsymbol{\lambda}, \boldsymbol{\mu}) \leq f(\mathbf{x}')$ for all $\mathbf{x}' \in \mathcal{F}$, it follows that $q(\boldsymbol{\lambda}, \boldsymbol{\mu}) \leq f^*$.

However, the lower bound $q(\lambda, \mu) \leq f^*$ is nontrivial only when $q(\lambda, \mu) > -\infty$. A natural question that arises is to find the best nontrivial lower bound on $f^*$. This motivates the definition of the *dual problem* as shown below.

$$\boxed{\begin{array}{l} \text{maximize } q(\lambda, \mu) \\ \text{subject to } \lambda \geq 0 \end{array}} \tag{2.12}$$

A *dual feasible solution* is a point $(\lambda, \mu) \in \mathbb{R}^{L+M}$ such that $\lambda \geq 0$ and $(\lambda, \mu) \in \mathcal{D}_q$, where $\mathcal{D}_q$ is the domain set of $q$, i.e.

$$\mathcal{D}_q = \{(\lambda, \mu)|q(\lambda, \mu) > -\infty\}.$$

For convenience, let $\mathcal{F}_d$ denote the feasible set for the dual problem. One interesting fact is that the dual problem is always a convex optimization problem even if the primal problem is not. The following theorem states this property formally.

**Theorem 2.8 (Convexity of the dual problem)** Consider an optimization problem of the form in (2.7) but without the assumption that $f$ is a convex function and $\mathcal{F}$ is a convex set. The corresponding dual problem in (2.12) is always a convex optimization problem.

**Proof** For any $(\lambda, \mu)$, $(\lambda', \mu') \in \mathcal{F}_d$ and $\alpha \in [0, 1]$, it is straightforward to write

$$\Lambda(\mathbf{x}, \alpha\lambda + (1-\alpha)\lambda', \alpha\mu + (1-\alpha)\mu')$$

$$= \alpha\Lambda(\mathbf{x}, \lambda, \mu) + (1-\alpha)\Lambda(\mathbf{x}, \lambda', \mu').$$

It follows that

$$\inf_{\mathbf{x} \in \mathcal{X}} \Lambda(\mathbf{x}, \alpha\lambda + (1-\alpha)\lambda', \alpha\mu + (1-\alpha)\mu')$$

$$\geq \alpha \inf_{\mathbf{x} \in \mathcal{X}} \Lambda(\mathbf{x}, \lambda, \mu) + (1-\alpha) \inf_{\mathbf{x} \in \mathcal{X}} \Lambda(\mathbf{x}, \lambda', \mu'),$$

yielding

$$q(\mathbf{x}, \alpha\lambda + (1-\alpha)\lambda', \alpha\mu + (1-\alpha)\mu')$$

$$\geq \alpha q(\mathbf{x}, \lambda, \mu) + (1-\alpha) q(\mathbf{x}, \lambda', \mu').$$

The last inequality implies that if $(\lambda, \mu)$, $(\lambda', \mu') \in \mathcal{F}_d$, i.e. $\lambda, \lambda' \geq 0$ and $q(\lambda, \mu)$, $q(\lambda', \mu') \geq -\infty$, then so is their convex combination. Thus, $\mathcal{F}_d$ is a convex set.

In addition, the same inequality implies the concavity of $q$. Thus, maximizing $q$, which is equivalent to minimizing the convex function $- q$, over the convex set $\mathcal{F}_d$ is a convex optimization problem.    □

Since any dual feasible solution $(\lambda, \mu) \in \mathcal{F}_d$ yields a lower bound $q(\lambda, \mu) \leq f^*$, it follows that the supremum of the dual function, denoted by

$$q^* = \sup_{\lambda \geq 0,\, \mu \in \mathbb{R}^M} q(\lambda, \mu), \tag{2.13}$$

is also a lower bound on $f^*$. This statement is known as the *weak duality theorem* and is stated formally below.

**Theorem 2.9 (Weak duality theorem)**    Let $f^*$ be the optimal cost of the primal problem in (2.7) and $q^*$ be the supremum of the dual function as defined in (2.13). Then, $q^* \leq f^*$.

Note that $q^*$ is equal to the dual optimal cost when it exists. The difference $f^* - q^*$ is called the *duality gap*. In some problems, the duality gap is zero. Such cases can be described by stating that *strong duality* holds, i.e. $f^* = q^*$. With strong duality, the optimal cost $f^*$ can be obtained from the dual problem.

## 2.5    Lagrange Multipliers

Dual variables $(\lambda^*, \mu^*) \in \mathcal{F}_d$ are *Lagrange multipliers* for the primal problem if $\lambda^* \geq 0$ and

$$f^* = \inf_{\mathbf{x} \in \mathcal{X}} \Lambda(\mathbf{x}, \lambda^*, \mu^*) = q(\lambda^*, \mu^*). \tag{2.14}$$

In general, Lagrange multipliers may or may not exist. When they exist, they may not be unique. When Lagrange multipliers exist, strong duality holds, i.e. $f^* = q^*$. To see this, use (2.13) and (2.14) to write $f^* = q(\lambda^*, \mu^*) \leq q^*$. From weak duality, i.e. $q^* \leq f^*$, it follows that $f^* = q^*$.

Note that a zero duality gap does not imply the existence of Lagrange multipliers when there is no dual optimal solution, as will be illustrated shortly by one example below. However, if the duality gap is zero and there is a dual optimal solution, then a dual optimal solution is a set of Lagrange multipliers. Below are some example scenarios of Lagrange multipliers.

**Example 2.3:**  Consider minimizing $f(x) = (x + 1)^2$ subject to $x \geq 0$. The Lagrangian for this problem is $\Lambda(x, \lambda) = (x + 1)^2 - \lambda x$. The corresponding dual function is equal to

$$q(\lambda) = \inf_{x \in \mathbb{R}} (x + 1)^2 - \lambda x = 1 - \frac{(2 - \lambda)^2}{4},$$

and is illustrated in figure 2.10. Note that the dual function is concave with the maximum equal to 1. In addition, since the primal optimal cost is $f^* = 1$, it is clear from the figure that $q(\lambda)$ is a lower bound of $f^*$.

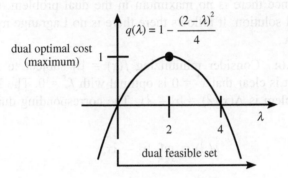

**Figure 2.10**  Illustration of the dual function $q(\lambda) = 1 - (2 - \lambda)^2/4$.

The dual problem is to maximize $q(\lambda) = 1 - (2 - \lambda)^2/4$ subject to $\lambda \geq 0$. By inspection, the dual optimal solution is $\lambda^* = 2$, with the dual optimal cost $q^* = 1$. Since $q^* = f^*$, there is no duality gap.

In summary, this problem has a unique primal optimal solution and a unique dual optimal solution with no duality gap. It follows that there is a unique Lagrange multiplier, which is $\lambda^* = 2$.  □

**Example 2.4:**  Consider minimizing $f(\mathbf{x}) = x_1 + x_2$ subject to $x_1 \geq 0$. The Lagrangian for this problem is $\Lambda(\mathbf{x}, \lambda) = x_1 + x_2 - \lambda x_1 = (1 - \lambda)x_1 + x_2$. The corresponding dual function is

$$q(\lambda) = \inf_{\mathbf{x} \in \mathbb{R}^2} (1 - \lambda)x_1 + x_2,$$

which is equal to $-\infty$. It follows that $\mathcal{D}_q = \emptyset$. Hence, the dual problem is infeasible. Since a Lagrange multiplier is equal to a dual optimal solution, it follows that there is no Lagrange multiplier in this problem.  □

**Example 2.5 (from [Bertsekas, 1995]):**   Consider minimizing $f(x) = x$ subject to $x^2 \leq 0$. Observe that $x = 0$ is the only feasible solution. Thus, $x^* = 0$ is optimal with $f^* = 0$. The Lagrangian for this problem is $\Lambda(x, \lambda) = x + \lambda x^2$. The corresponding dual function is

$$q(\lambda) = \inf_{x \in \mathbb{R}} x + \lambda x^2 = -\frac{1}{4\lambda}.$$

Note that $\mathcal{D}_q = (0, \infty)$. The dual problem is then to maximize $-1/4\lambda$ subject to $\lambda > 0$. Since $q^* = \sup_{\lambda > 0} -1/4\lambda = 0$, there is no duality gap. However, since there is no maximum in the dual problem, there is no dual optimal solution. It follows there there is no Lagrange multiplier in this problem.                                        □

**Example 2.6:**   Consider minimizing $f(x) = |x|$ subject to $x \geq 0$. By inspection, it is clear that $x^* = 0$ is optimal with $f^* = 0$. The Lagrangian for this problem is $\Lambda(x, \lambda) = |x| - \lambda x$. The corresponding dual function is

$$q(\lambda) = \inf_{x \in \mathbb{R}} |x| - \lambda x,$$

which is equal to 0 for $\lambda \in [-1, 1]$ and is undefined for $\lambda \notin [-1, 1]$, as illustrated in figure 2.11. From the constraint $\lambda \geq 0$ in the dual problem, it follows that any $\lambda^* \in [0, 1]$ is dual optimal with $q^* = 0$. Since $q^* = f^*$, there is no duality gap.

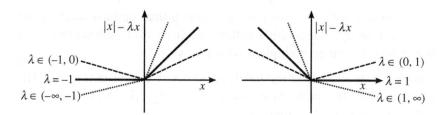

**Figure 2.11**   Illustration of the function $|x| - \lambda x$.

In conclusion, since any $\lambda^* \in [0, 1]$ is dual optimal with no duality gap, there are infinitely many Lagrange multipliers for this problem. □

## Slator Conditions for Existence of Lagrange Multipliers

One simple set of conditions called the *Slator conditions* guarantee the existence of at least one set of Lagrange multipliers for the convex optimization problem in (2.7). The Slator conditions are given below [Bertsekas, 1995].

1. $\mathcal{X}$ is a convex set.
2. $f, g_1, ..., g_L$ are convex functions.
3. $h_1, ..., h_M$ are such that $\mathbf{h}(\mathbf{x})$ can be expressed as $\mathbf{Ax} + \mathbf{b}$ for some matrix $\mathbf{A}$ and vector $\mathbf{b}$. In addition, $\mathbf{A}$ has full rank, i.e. $\text{rank}(\mathbf{A}) = M$.
4. The primal optimal cost $f^*$ is finite.
5. There exists a feasible solution $\mathbf{x}' \in \mathcal{X}$ such that $\mathbf{g}(\mathbf{x}') < \mathbf{0}$.

Roughly speaking, the last statement says that there is at least one feasible point in the strict interior of the feasible set $\mathcal{F}$. The following theorem formally states that the Slator conditions guarantee the existence of Lagrange multipliers.

**Theorem 2.10 (Strong duality for Slator conditions)** For the convex optimization problem in (2.7), if the Slator conditions are satisfied, then there exists a set of Lagrange multipliers.

†**Proof** For convenience, define vectors $\mathbf{v} = (v_1, ..., v_L)$ and $\mathbf{w} = (w_1, ..., w_M)$. Consider the set

$\mathcal{A} = \{(\mathbf{v}, \mathbf{w}, u)|$ there is $\mathbf{x} \in \mathcal{X}$ such that $\mathbf{g}(\mathbf{x}) \le \mathbf{v}$, $\mathbf{h}(\mathbf{x}) = \mathbf{w}$, and $f(\mathbf{x}) \le u\}$

Notice that, for any $\mathbf{x} \in \mathcal{X}$, $(\mathbf{g}(\mathbf{x}), \mathbf{h}(\mathbf{x}), f(\mathbf{x}))$ is in $\mathcal{A}$. First, it will be shown that $\mathcal{A}$ is convex. Let $(\mathbf{v}, \mathbf{w}, u), (\mathbf{v}', \mathbf{w}', u') \in \mathcal{A}$. From the definition of $\mathcal{A}$, there exist $\mathbf{x}, \mathbf{x}' \in \mathcal{X}$ such that

$\mathbf{g}(\mathbf{x}) \le \mathbf{v}$, $\mathbf{h}(\mathbf{x}) = \mathbf{w}$, $f(\mathbf{x}) \le u$, and $\mathbf{g}(\mathbf{x}') \le \mathbf{v}'$, $\mathbf{h}(\mathbf{x}') = \mathbf{w}'$, $f(\mathbf{x}') \le u'$.

Let $\alpha \in [0, 1]$. From the convexity of $f, g_1, ..., g_L$, and the form of $h_1, ..., h_M$,

$$\mathbf{g}(\alpha\mathbf{x} + (1 - \alpha)\mathbf{x}') \le \alpha\mathbf{g}(\mathbf{x}) + (1 - \alpha)\,\mathbf{g}(\mathbf{x}') \le \alpha\mathbf{v} + (1 - \alpha)\mathbf{v}',$$

$$\mathbf{h}(\alpha\mathbf{x} + (1 - \alpha)\mathbf{x}') = \alpha\mathbf{h}(\mathbf{x}) + (1 - \alpha)\,\mathbf{h}(\mathbf{x}') = \alpha\mathbf{w} + (1 - \alpha)\mathbf{w}',$$

$$f(\alpha\mathbf{x} + (1 - \alpha)\mathbf{x}') \le \alpha f(\mathbf{x}) + (1 - \alpha)\,f(\mathbf{x}') \le \alpha u + (1 - \alpha)u'.$$

Since $\alpha\mathbf{x} + (1 - \alpha)\mathbf{x}' \in \mathcal{X}$, it follows that $\alpha(\mathbf{v}, \mathbf{w}, u) + (1 - \alpha)(\mathbf{v}', \mathbf{w}', u') \in \mathcal{A}$, proving the convexity of $\mathcal{A}$.

It is next argued by contradiction that $(0, 0, f^*)$ is not an interior point of $\mathcal{A}$. Suppose that $(0, 0, f^*)$ is an interior point of $\mathcal{A}$. Then, for some sufficiently small $\varepsilon > 0$, $(0, 0, f^* - \varepsilon) \in \mathcal{A}$, contradicting the assumption that $f^*$ is the primal optimal cost.

Since $(0, 0, f^*)$ is not an interior point of $\mathcal{A}$, the supporting hyperplane theorem (see theorem 2.6) implies that there exists a hyperplane that passes through $(0, 0, f^*)$ and supports $\mathcal{A}$. This hyperplane is described by parameters $\mathbf{b} \in \mathbb{R}^L$, $\mathbf{c} \in \mathbb{R}^M$, $a \in \mathbb{R}$ with $(\mathbf{b}, \mathbf{c}, a) \neq (\mathbf{0}, \mathbf{0}, 0)$ such that

$$af^* \leq \mathbf{b}^T\mathbf{v} + \mathbf{c}^T\mathbf{w} + au \quad \text{for all} \quad (\mathbf{v}, \mathbf{w}, u) \in \mathcal{A}. \tag{2.15}$$

Let $\mathbf{e}_l \in \mathbb{R}^L$ denote the unit vector with the $l$th component equal to 1. From the definition of $\mathcal{A}$, if $(\mathbf{v}, \mathbf{w}, u) \in \mathcal{A}$, so are $(\mathbf{v}, \mathbf{w}, u + \varepsilon)$ and $(\mathbf{v} + \varepsilon\mathbf{e}_l, \mathbf{w}, u)$ for all $l \in \{1, ..., L\}$. Substituting $(\mathbf{v}, \mathbf{w}, u)$ and $(\mathbf{v}, \mathbf{w}, u + \varepsilon)$ in (2.15) and subtracting the two inequalities yields $a \geq 0$. A similar procedure with $(\mathbf{v}, \mathbf{w}, u)$ and $(\mathbf{v} + \varepsilon\mathbf{e}_l, \mathbf{w}, u)$ yields $\mathbf{b} \geq \mathbf{0}$.

It is next shown by contradiction that $a > 0$. Suppose that $a = \mathbf{0}$. Then, (2.15) yields $0 \leq \mathbf{b}^T\mathbf{v} + \mathbf{c}^T\mathbf{w}$ for all $(\mathbf{v}, \mathbf{w}, u) \in \mathcal{A}$. Recall that, by assumption, there is a feasible solution $\mathbf{x}' \in \mathcal{X}$ with $\mathbf{g}(\mathbf{x}') < \mathbf{0}$. Since $(\mathbf{g}(\mathbf{x}'), \mathbf{h}(\mathbf{x}'), f(\mathbf{x}')) \in \mathcal{A}$,

$$0 \leq \mathbf{b}^T\mathbf{g}(\mathbf{x}') + \mathbf{c}^T\mathbf{h}(\mathbf{x}') = \mathbf{b}^T\mathbf{g}(\mathbf{x}'),$$

where the last equality follows from the feasibility of $\mathbf{x}'$, i.e. $\mathbf{h}(\mathbf{x}') = \mathbf{0}$. From $\mathbf{g}(\mathbf{x}') < \mathbf{0}$ and $\mathbf{b} \geq \mathbf{0}$, it follows that $\mathbf{b} = \mathbf{0}$, yielding $\mathbf{c} \neq \mathbf{0}$ since $(\mathbf{b}, \mathbf{c}, a) \neq (\mathbf{0}, \mathbf{0}, 0)$. The condition in (2.15) is then simplied to $0 \leq \mathbf{c}^T\mathbf{w}$ for all $(\mathbf{v}, \mathbf{w}, u) \in \mathcal{A}$.

Since $\mathbf{x}'$ is in the interior of the feasible set, the fact that $0 = \mathbf{c}^T\mathbf{h}(\mathbf{x}')$ $= \mathbf{c}^T(\mathbf{A}\mathbf{x}' + \mathbf{b}) = \mathbf{c}^T\mathbf{A}\mathbf{x}'$ implies that there exists a feasible solution $\mathbf{x} \in \mathcal{X}$ such that $\mathbf{c}^T\mathbf{h}(\mathbf{x}) = \mathbf{c}^T\mathbf{A}\mathbf{x} < 0$ unless $\mathbf{c}^T\mathbf{A} = \mathbf{0}$. The condition $0 \leq \mathbf{c}^T\mathbf{w}$ for $\mathbf{w} = \mathbf{h}(\mathbf{x})$ implies that $\mathbf{c}^T\mathbf{A} = \mathbf{0}$. However, the condition $\mathbf{c}^T\mathbf{A} = \mathbf{0}$ together with $\mathbf{c} \neq \mathbf{0}$ contradicts the assumption that $\mathbf{A}$ has full rank.

In summary, $a > 0$ and $\mathbf{b} \geq \mathbf{0}$. Without loss of generality, $(\mathbf{b}, \mathbf{c}, a)$ can be normalized so that $a = 1$. With $a = 1$, (2.15) yields

$$f^* \leq \mathbf{b}^T\mathbf{g}(\mathbf{x}) + \mathbf{c}^T\mathbf{h}(\mathbf{x}) + f(\mathbf{x}) \text{ for all } \mathbf{x} \in \mathcal{X}.$$

Taking the infimum over $\mathbf{x} \in \mathcal{X}$ and noting that $\mathbf{b} \geq \mathbf{0}$,

$$f^* \leq \inf_{\mathbf{x} \in \mathcal{X}} \ [\mathbf{b}^\mathrm{T}\mathbf{g}(\mathbf{x}) + \mathbf{c}^\mathrm{T}\mathbf{h}(\mathbf{x}) + f(\mathbf{x})]$$

$$= q(\mathbf{b}, \mathbf{c}) \leq \sup_{\mathbf{b} \geq 0, \, \mathbf{c} \in \mathbb{R}^M} q(\mathbf{b}, \mathbf{c}) = q^*.$$

Using weak duality, i.e. $q \leq f^*$, it follows that

$$f^* = \inf_{\mathbf{x} \in \mathcal{X}} \ [\mathbf{b}^\mathrm{T}\mathbf{g}(\mathbf{x}) + \mathbf{c}^\mathrm{T}\mathbf{h}(\mathbf{x}) + f(\mathbf{x})] = q(\mathbf{b}, \mathbf{c}) = q^*,$$

which implies that $(\mathbf{b}, \mathbf{c})$ is a dual optimal solution with no duality gap. Hence, $\mathbf{b}$ and $\mathbf{c}$ are Lagrange multipliers for the convex optimization problem in (2.7). □

## 2.6   Primal-Dual Optimality Conditions

Consider the convex optimization problem in (2.7) under an additional assumption that Lagrange multipliers exist. The definition of Lagrange multipliers yields the following optimality condition.

**Theorem 2.11 (Primal-dual optimality conditions)** For the convex optimization problem in (2.7), $(\mathbf{x}^*, \boldsymbol{\lambda}^*, \boldsymbol{\mu}^*)$ is a primal-dual optimal solution pair if and only if the following conditions hold.

1. Primal feasibility: $\mathbf{g}(\mathbf{x}^*) \leq \mathbf{0}$ and $\mathbf{h}(\mathbf{x}^*) = \mathbf{0}$
2. Dual feasibility: $\boldsymbol{\lambda}^* \geq \mathbf{0}$
3. Lagrangian optimality: $\mathbf{x}^* = \arg\min_{\mathbf{x} \in \mathcal{X}} \Lambda(\mathbf{x}, \boldsymbol{\lambda}^*, \boldsymbol{\mu}^*)$
4. Complimentary slackness: $\lambda_l^* g_l(\mathbf{x}^*) = 0$ for all $l \in \{1, ..., L\}$

**Proof**   Suppose that $(\mathbf{x}^*, \boldsymbol{\lambda}^*, \boldsymbol{\mu}^*)$ is a primal-dual optimal solution pair. From the feasibility of $(\mathbf{x}^*, \boldsymbol{\lambda}^*, \boldsymbol{\mu}^*)$, statements 1 and 2 hold. To justify statements 3 and 4, write

$$f^* = f(\mathbf{x}^*) \geq f(\mathbf{x}^*) + \underbrace{\boldsymbol{\lambda}^{*\mathrm{T}}\mathbf{g}(\mathbf{x}^*)}_{\leq 0} + \underbrace{\boldsymbol{\mu}^{*\mathrm{T}}\mathbf{h}(\mathbf{x}^*)}_{= 0}$$

$$= \Lambda(\mathbf{x}^*, \boldsymbol{\lambda}^*, \boldsymbol{\mu}^*) \geq \inf_{\mathbf{x} \in \mathcal{X}} \Lambda(\mathbf{x}, \boldsymbol{\lambda}^*, \boldsymbol{\mu}^*),$$

where the first inequality follows from statements 1 and 2, i.e. $\boldsymbol{\lambda}^* \geq \mathbf{0}$, $\mathbf{g}(\mathbf{x}^*) \leq \mathbf{0}$, and $\mathbf{h}(\mathbf{x}^*) = \mathbf{0}$. Since $(\boldsymbol{\lambda}^*, \boldsymbol{\mu}^*)$ is a set of Lagrange multipliers, $f^* = \inf_{\mathbf{x} \in \mathcal{X}} \Lambda(\mathbf{x}, \boldsymbol{\lambda}^*, \boldsymbol{\mu}^*)$. It follows that

$$f^* = f(\mathbf{x}^*) = f(\mathbf{x}^*) + \boldsymbol{\lambda}^{*\mathrm{T}}\mathbf{g}(\mathbf{x}^*) + \boldsymbol{\mu}^{*\mathrm{T}}\mathbf{h}(\mathbf{x}^*)$$

$$= \inf_{\mathbf{x}\in\mathcal{X}} \Lambda(\mathbf{x}, \boldsymbol{\lambda}^*, \boldsymbol{\mu}^*),$$

which implies statements 3 and 4. More specifically, for statement 3, the fact that $f(\mathbf{x}^*) = \inf_{\mathbf{x}\in\mathcal{X}} \Lambda(\mathbf{x}, \boldsymbol{\lambda}^*, \boldsymbol{\mu}^*)$ implies that $\mathbf{x}^* = \arg\min_{\mathbf{x}\in\mathcal{X}} \Lambda(\mathbf{x}, \boldsymbol{\lambda}^*, \boldsymbol{\mu}^*)$. In addition, for statement 4, having $\boldsymbol{\lambda}^{*\mathrm{T}}\mathbf{g}(\mathbf{x}^*) = 0$ and $\boldsymbol{\lambda}^* \geq \mathbf{0}$ is equivalent to having

$$\lambda_l^* g_l(\mathbf{x}^*) = 0 \quad \text{for all } l \in \{1, ..., L\}.$$

For the converse, suppose that $(\mathbf{x}^*, \boldsymbol{\lambda}^*, \boldsymbol{\mu}^*)$ satisfies all four statements. Then,

$$f^* \leq f(\mathbf{x}^*) = \Lambda(\mathbf{x}^*, \boldsymbol{\lambda}^*, \boldsymbol{\mu}^*) = \min_{\mathbf{x}\in\mathcal{X}} \Lambda(\mathbf{x}, \boldsymbol{\lambda}^*, \boldsymbol{\mu}^*)$$

$$= q(\boldsymbol{\lambda}^*, \boldsymbol{\mu}^*) \leq q^*.$$

Together with weak duality, i.e. $q^* \leq f^*$, it follows that $f^* = q^*$. Since there is no duality gap, $f^* = f(\mathbf{x}^*)$ and $q^* = q(\boldsymbol{\lambda}^*, \boldsymbol{\mu}^*)$. In other words, $(\mathbf{x}^*, \boldsymbol{\lambda}^*, \boldsymbol{\mu}^*)$ is a primal-dual optimal solution pair.    □

Statement 4 of theorem 2.11 is referred to as *complimentary slackness*, which indicates that the Lagrange multiplier $\lambda_l^*$ for a constraint that is not *active* at optimal solution $\mathbf{x}^*$, i.e. $g_l(\mathbf{x}^*) < 0$, must be zero.

Assume now that the functions $f$, $g_l$, and $h_m$ are differentiable in $\mathcal{X}$. It follows that statement 3 of theorem 2.11 can be replaced by the derivative condition below.

$$\nabla f(\mathbf{x}^*) + \sum_{l=1}^{L}\lambda_l^*\nabla g_l(\mathbf{x}^*) + \sum_{m=1}^{M}\mu_m^*\nabla h_m(\mathbf{x}^*) = \mathbf{0} \qquad (2.16)$$

The condition in (2.16) together with statements 1, 2, and 4 of theorem 2.11 yield the *Karush-Kuhn-Tucker (KKT) conditions* for optimality, as stated formally in the following theorem.

**Theorem 2.12 (KKT conditions)** For the convex optimization problem in (2.7) where $f$, $g_l$, and $h_m$ are differentiable in $\mathcal{X}$, $(\mathbf{x}^*, \boldsymbol{\lambda}^*, \boldsymbol{\mu}^*)$ is a primal-dual optimal solution pair if and only if the following conditions hold.

1. Primal feasibility: $\mathbf{g}(\mathbf{x}^*) \leq \mathbf{0}$ and $\mathbf{h}(\mathbf{x}^*) = \mathbf{0}$
2. Dual feasibility: $\lambda^* \geq \mathbf{0}$
3. Lagrangian optimality: $\nabla f(\mathbf{x}^*) + \sum_{l=1}^{L} \lambda_l^* \nabla g_l(\mathbf{x}^*) + \sum_{m=1}^{M} \mu_m^*$
   $\nabla h_m(\mathbf{x}^*) = \mathbf{0}$
4. Complimentary slackness: $\lambda_l^* g_l(\mathbf{x}^*) = 0$ for all $l \in \{1, ..., L\}$

In some cases, the KKT conditions can be used to solve the problem analytically, yielding both primal and dual optimal solutions as closed form expressions. In other cases, numerical algorithms may be developed based on the KKT conditions. In particular, several algorithms for convex optimization can be thought of as numerical methods for solving the KKT conditions [Boyd and Vandenberghe, 2004]. The following simple example illustrates how KKT conditions are used to derive a primal-dual optimal solution pair.

**Example 2.7:** Consider minimizing $f(x) = (x + 1)^2$ subject to $x \geq 0$ using the KKT conditions. The Lagrangian for this problem is

$$\Lambda(x, \lambda) = (x + 1)^2 - \lambda x.$$

Let $(x^*, \lambda^*)$ be a primal-dual optimal solution pair. From Lagrangian optimality,

$$2(x^* + 1) - \lambda^* = 0.$$

From complimentary slackness, $\lambda^* x^* = 0$. First, suppose that $x^* \neq 0$. Then, $\lambda^* = 0$, yielding $x^* = -1$ which is not possible since it is not primal feasible.

Suppose now that $\lambda^* \neq 0$. Then, $x^* = 0$, yielding $\lambda^* = 2$ which is dual feasible. It follows that $(x^*, \lambda^*) = (0, 2)$, consistent with examples 2.1 and 2.3. □

## 2.7 Sensitivity Analysis

Consider a convex optimization problem in (2.7) with Lagrange multipliers. This section discusses how Lagrange multipliers are useful information about the sensitivity of the optimal cost with respect to perturbations of the constraints. The following discussion is adapted from [Boyd and Vandenberghe, 2004].

Consider the following perturbed problem, which is a slight modification of (2.7).

$$\text{minimize } f(\mathbf{x})$$

$$\text{subject to } \forall l \in \{1, \dots, L\}, g_l(\mathbf{x}) \le v_l$$

$$\forall m \in \{1, \dots, M\}, h_m(\mathbf{x}) = w_m \qquad (2.17)$$

Let $\mathbf{v} = (v_1, \dots, v_L)$ and $\mathbf{w} = (w_1, \dots, w_M)$. Define $f^*(\mathbf{v}, \mathbf{w})$ as the optimal cost for the perturbed problem. Note that $f^*(\mathbf{0}, \mathbf{0}) = f^*$. The following lemma establishes a useful inequality.

**Lemma 2.2:** Let $(\boldsymbol{\lambda}^*, \boldsymbol{\mu}^*)$ be a Lagrange multiplier vector. Then, for all $(\mathbf{v}, \mathbf{w})$,

$$f^*(\mathbf{v}, \mathbf{w}) \ge f^*(\mathbf{0}, \mathbf{0}) - \boldsymbol{\lambda}^{*\mathrm{T}}\mathbf{v} - \boldsymbol{\mu}^{*\mathrm{T}}\mathbf{w}.$$

**Proof** Let $\mathbf{x}$ be any feasible solution to the perturbed problem, i.e. $\mathbf{g}(\mathbf{x}) \le \mathbf{v}$ and $\mathbf{h}(\mathbf{x}) = \mathbf{w}$. From strong duality, the definition of the dual function, and the feasibility of $\mathbf{x}$ in the perturbed problem,

$$f^*(\mathbf{0}, \mathbf{0}) = q(\boldsymbol{\lambda}^*, \boldsymbol{\mu}^*) \le f(\mathbf{x}) + \boldsymbol{\lambda}^{*\mathrm{T}}\mathbf{g}(\mathbf{x}) + \boldsymbol{\mu}^{*\mathrm{T}}\mathbf{h}(\mathbf{x})$$

$$\le f(\mathbf{x}) + \boldsymbol{\lambda}^{*\mathrm{T}}\mathbf{v} + \boldsymbol{\mu}^{*\mathrm{T}}\mathbf{w}.$$

The above inequality can be written as $f(\mathbf{x}) \ge f^*(\mathbf{0}, \mathbf{0}) - \boldsymbol{\lambda}^{*\mathrm{T}}\mathbf{v} - \boldsymbol{\mu}^{*\mathrm{T}}\mathbf{w}$. Since the inequality holds for any feasible $\mathbf{x}$ in the perturbed problem, it follows that $f^*(\mathbf{v}, \mathbf{w}) \ge f^*(\mathbf{0}, \mathbf{0}) - \boldsymbol{\lambda}^{*\mathrm{T}}\mathbf{v} - \boldsymbol{\mu}^{*\mathrm{T}}\mathbf{w}.$  □

The additional assumption that $f^*(\mathbf{v}, \mathbf{w})$ is differentiable at $(\mathbf{v}, \mathbf{w}) = (\mathbf{0}, \mathbf{0})$ yields the following theorem.

**Theorem 2.13 (Local sensitivity)** If $f^*(\mathbf{v}, \mathbf{w})$ is differentiable at $(\mathbf{v}, \mathbf{w}) = (\mathbf{0}, \mathbf{0})$, then

$$\lambda_l^* = -\frac{\partial f^*(\mathbf{0}, \mathbf{0})}{\partial v_l}, \, l \in \{1, \dots, L\},$$

$$\mu_m^* = -\frac{\partial f^*(\mathbf{0}, \mathbf{0})}{\partial w_m}, \, m \in \{1, \dots, M\}.$$

**Proof** Consider first the perturbation $(\mathbf{v}, \mathbf{w}) = (t\mathbf{e}_l, \mathbf{0})$, where $t \in \mathbb{R}$ is small and $\mathbf{e}_l \in \mathbb{R}^L$ is the unit vector whose $l$th component is equal to 1. From the definition of differentiation,

$$\frac{\partial f^*(\mathbf{0}, \mathbf{0})}{\partial v_l} = \lim_{t \to 0} \frac{f^*(t\mathbf{e}_l, \mathbf{0}) - f^*}{t}.$$

From the inequality in lemma 2.2,

$$\frac{f^*(t\mathbf{e}_l, \mathbf{0}) - f^*}{t} \ge -\lambda_l^* \quad \text{for } t > 0,$$

$$\frac{f^*(t\mathbf{e}_l, \mathbf{0}) - f^*}{t} \le -\lambda_l^* \quad \text{for } t < 0.$$

Taking the limit as $t \to 0$ with $t > 0$ on one hand and with $t < 0$ on the other yields

$$-\lambda_l^* \le \frac{\partial f^*(\mathbf{0}, \mathbf{0})}{\partial v_l} \le -\lambda_l^*,$$

which implies that $\dfrac{\partial f^*(\mathbf{0}, \mathbf{0})}{\partial v_l} = -\lambda_l^*$. Proving that $\dfrac{\partial f^*(\mathbf{0}, \mathbf{0})}{\partial w_m} = -\mu_m^*$ can be done using similar arguments and is thus omitted. □

The local sensitivity specifies how sensitive the optimal cost is to each constraint perturbation at optimal solution $\mathbf{x}^*$. An interesting special case is when $g_l(\mathbf{x}^*) < 0$, i.e. the constraint is not active. By complimentary slackness, the corresponding Lagrange multiplier $\lambda_l^*$ must be zero. This means that the constraint can be further tightened without affecting the optimal cost.

On the other hand, suppose that $g_l(\mathbf{x}^*) = 0$, i.e. the constraint is active. Then, the Lagrange multiplier $\lambda_l^*$ specifies the change in the optimal cost associated with the amount of constraint perturbation. The larger the magnitude of $\lambda_l^*$, the larger the optimal cost changes when the constraint is perturbed by a fixed amount.

**Example 2.8:** Let $\mathbf{x} \in \mathbb{R}^2$. Consider minimizing $f(\mathbf{x}) = (2x_1 + 1)^2 + (x_2 + 1)^2$ subject to $\mathbf{x} \ge \mathbf{0}$. The Lagrangian for this problem is

$$\Lambda(\mathbf{x}, \lambda) = (2x_1 + 1)^2 + (x_2 + 1)^2 - \lambda_1 x_1 - \lambda_2 x_2.$$

Let $(\mathbf{x}^*, \lambda^*)$ be a primal-dual optimal solution pair. The KKT conditions require that

$$\lambda_1^* = 4(2x_1^* + 1), \quad \lambda_1^* x_1^* = 0,$$
$$\lambda_2^* = 2(x_2^* + 1), \quad \lambda_2^* x_2^* = 0.$$

First, suppose that $x_1^* \neq 0$. Then, $\lambda_1^* = 0$, yielding $x_1^* = -1/2$ which is not primal feasible. Suppose now that $\lambda_1^* \neq 0$. Then, $x_1^* = 0$, yielding $\lambda_1^* = 4$ which is dual feasible. Hence, $x_1^* = 0$ and $\lambda_1^* = 4$. Similar arguments yield $x_2^* = 0$ and $\lambda_2^* = 2$. Accordingly, the primal optimal cost is $f^* = 2$. The contour lines together with the primal optimal solution are illustrated in figure 2.12.

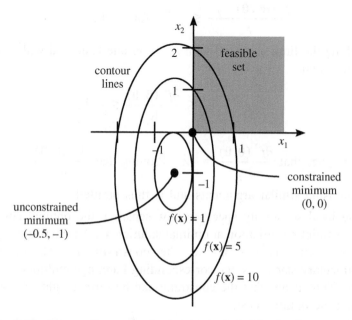

**Figure 2.12**   Contour lines and primal optimal solution for the demonstration of local sensitivity.

Since $\lambda_1^* > \lambda_2^*$, one can expect that relaxing (tightening) the constraint $x_1 \geq 0$ will affect the primal optimal cost more than relaxing (tightening) the constraint $x_2 \geq 0$. In particular, if $x_1 \geq 0$ is relaxed to $x_1 \geq -\varepsilon$ for some small $\varepsilon > 0$, then the primal optimal solution becomes (by inspection of figure 2.12) $\mathbf{x}^* = (-\varepsilon, 0)$ with the optimal cost

$$f^* = (-2\varepsilon + 1)^2 + 1 \approx 2 - 4\varepsilon.$$

On the other hand, if $x_2 \geq 0$ is relaxed to $x_2 \geq -\varepsilon$, then $\mathbf{x}^* = (0, -\varepsilon)$ with the optimal cost

$$f^* = 1 + (-\varepsilon + 1)^2 \approx 2 - 2\varepsilon.$$

Notice how the changes of $f^*$ per unit of constraint perturbation are equal to 4 and 2, which are the values of the Lagrange multipliers $\lambda_1^*$ and $\lambda_2^*$ respectively. □

## 2.8 Notes on Maximization Problems

While a maximization problem can always be transformed into a minimization problem, it is convenient to rewrite various statements, e.g. KKT conditions, in alternative forms that can directly be applied to a maximization problem. In the discussion below, statements will be made without justification since the arguments are identical to those for minimization problems.

Consider the following maximization problem which differs from the problem in (2.7) in two aspects. First, the objective is to maximize instead of to minimize. Second, each inequality constraint has the form $g_l(\mathbf{x}) \geq 0$ instead of $g_l(\mathbf{x}) \leq 0$.

$$
\begin{aligned}
&\text{maximize } f(\mathbf{x})\\
&\text{subject to } \forall l \in \{1, ..., L\},\ g_l(\mathbf{x}) \geq 0\\
&\qquad\qquad \forall m \in \{1, ..., M\},\ h_m(\mathbf{x}) = 0
\end{aligned}
\tag{2.18}
$$

The Lagrangian for the optimization problem is still given by

$$
\Lambda(\mathbf{x}, \lambda, \mu) = f(\mathbf{x}) + \sum_{l=1}^{L} \lambda_l\, g_l(\mathbf{x}) + \sum_{m=1}^{M} \mu_m h_m(\mathbf{x})
$$

$$
= f(\mathbf{x}) + \lambda^{\mathrm{T}} \mathbf{g}(\mathbf{x}) + \mu^{\mathrm{T}} \mathbf{h}(\mathbf{x}),
$$

where $\lambda = (\lambda_1, ..., \lambda_L)$, $\mu = (\mu_1, ..., \mu_M)$, $\mathbf{g}(\mathbf{x}) = (g_1(\mathbf{x}), ..., g_L(\mathbf{x}))$, and $\mathbf{h}(\mathbf{x}) = (h_1(\mathbf{x}), ..., h_M(\mathbf{x}))$. The dual function is now defined as

$$
q(\lambda, \mu) = \sup_{\mathbf{x} \in \mathcal{X}} \Lambda(\mathbf{x}, \lambda, \mu),
\tag{2.19}
$$

where $\mathcal{X}$ is the domain set defined in (2.8).

Let $\mathbf{x}^*$ and $f^*$ denote an optimal solution and the optimal cost of the primal problem in (2.18). For any $\lambda \geq \mathbf{0}$ and $\mu \in \mathbb{R}^M$, the dual function in (2.19) provides an upper bound on $f^*$. Finding the best of such upper bounds leads to the dual problem defined as follows.

$$\boxed{\begin{array}{l} \text{minimize } q(\lambda, \mu) \\ \text{subject to } \lambda \geq 0 \end{array}} \qquad (2.20)$$

The infimum of the dual function, denoted by

$$q^* = \inf_{\lambda \geq 0, \mu \in \mathbb{R}^M} q(\lambda, \mu), \qquad (2.21)$$

is also an upper bound on $f^*$. The weak duality theorem is thus expressed as $q^* \geq f^*$. Accordingly, the duality gap is $q^* - f^*$.

Dual variables $(\lambda^*, \mu^*)$ are Lagrange multipliers for the primal problem if $\lambda^* \geq 0$ and

$$f^* = \sup_{x \in \mathcal{X}} \Lambda(x, \lambda^*, \mu^*) = q(\lambda^*, \mu^*). \qquad (2.22)$$

When the duality gap is zero and there is an optimal dual solution, an optimal dual solution yields a set of Lagrange multipliers.

Finally, when all the involved functions are differentiable, the KKT conditions can be expressed as follows.

1. Primal feasibility: $g(x^*) \geq 0$ and $h(x^*) = 0$
2. Dual feasibility: $\lambda^* \geq 0$
3. Lagrangian optimality: $\nabla f(x^*) + \sum_{l=1}^{L} \lambda_l^* \nabla g_l(x^*) + \sum_{m=1}^{M} \mu_m^* \nabla h_m(x^*) = 0$
4. Complimentary slackness: $\lambda_l^* g_l(x^*) = 0$ for all $l \in \{1, ..., L\}$

## 2.9 Numerical Algorithms for Unconstrained Optimization

Since only a limited number of convex optimization problems can be solved to obtain closed form solutions, it is practical to discuss algorithms for numerical solutions. Consider now a convex optimization problem with no constraint, i.e.

$$\text{minimize } f(x)$$
$$\text{subject to } x \in \mathbb{R}^N,$$

where $f$ is a convex function.

For unconstrained optimization, most numerical algorithms rely on the *iterative descent* approach, which involves moving from point $\mathbf{x}^k$ to point $\mathbf{x}^{k+1}$ in iteration $k$ such that $f(\mathbf{x}^{k+1}) < f(\mathbf{x}^k)$, where $k \in \mathbb{Z}^+$ and $\mathbf{x}^0$ is the initial point.[1] Figure 2.13 illustrates the iterative descent approach in $\mathbb{R}^2$. The key question is how to choose $\mathbf{x}^{k+1}$ given $\mathbf{x}^k$.

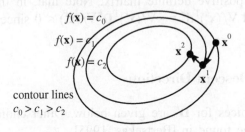

**Figure 2.13** Iterative descent in $\mathbb{R}^2$.

Consider moving from point $\mathbf{x}^k$ to point $\mathbf{x}^k + \alpha\mathbf{d}$, where $\alpha > 0$ is the *stepsize* and $\mathbf{d}$ is the *direction* vector. For small $\alpha > 0$, the first-order Taylor series approximation yields

$$f(\mathbf{x}^{k+1}) \approx f(\mathbf{x}^k) + \alpha\nabla f(\mathbf{x}^k)^{\mathrm{T}}\mathbf{d}.$$

To make $f(\mathbf{x}^{k+1}) < f(\mathbf{x}^k)$, $\mathbf{d}$ should be chosen such that $\nabla f(\mathbf{x}^k)^{\mathrm{T}}\mathbf{d} < 0$.[2] In other words, the angle between $\nabla f(\mathbf{x}^k)$ and $\mathbf{d}$ should be greater than 90 degree. However, $\alpha$ cannot be too large, as illustrated in figure 2.14.

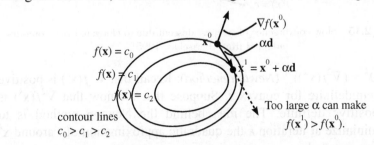

**Figure 2.14** Descent direction $\mathbf{d}$ with $\nabla f(\mathbf{x}^k)^{\mathrm{T}}\mathbf{d} < 0$.

The condition $\nabla f(\mathbf{x}^k)^{\mathrm{T}}\mathbf{d} < 0$ yields a class of iterations of the form

$$\mathbf{x}^{k+1} = \mathbf{x}^k + \alpha^k\mathbf{d}^k, \quad \text{where } \nabla f(\mathbf{x}^k)^{\mathrm{T}}\mathbf{d}^k < 0. \qquad (2.23)$$

---

[1] Let $\mathbb{Z}^+$ denote the set of nonnegative integers, i.e. $\mathbb{Z}^+ = \{0, 1, 2, ...\}$.
[2] Recall that $\nabla f(\mathbf{x}^k) = \mathbf{0}$ means that $\mathbf{x}^k$ is optimal. Hence, consider only nonzero $\nabla f(\mathbf{x}^k)$.

These iterations are referred to as *gradient methods*. In general, there is no single method of choosing $\alpha^k$ and $\mathbf{d}^k$ that can be recommended for all problems [Bertsekas, 1995]. Most gradient methods are of the form

$$\mathbf{x}^{k+1} = \mathbf{x}^k - \alpha^k \, \mathbf{D}^k \, \nabla f(\mathbf{x}^k), \tag{2.24}$$

where $\mathbf{D}^k$ is a positive definite matrix. Note that, in this form, $\mathbf{d}^k = -\mathbf{D}^k \nabla f(\mathbf{x}^k)$ and $\nabla f(\mathbf{x}^k)^{\mathrm{T}} \mathbf{d}^k = -\nabla f(\mathbf{x}^k)^{\mathrm{T}} \mathbf{D}^k \nabla f(\mathbf{x}^k) < 0$ since $\mathbf{D}^k$ is positive definite.

## Selection of Descent Direction

Two simple choices for $\mathbf{D}^k$ are given below. Other choices of descent directions can be found in [Bertsekas, 1995].

1. $\mathbf{D}^k = \mathbf{I}$ (*steepest descent*): In this case, $\mathbf{d}^k = -\nabla f(\mathbf{x}^k)$. Steepest descent is the simplest choice but can lead to slow convergence, as illustrated in figure 2.15.

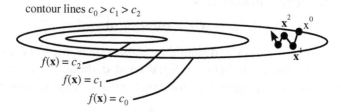

contour lines $c_0 > c_1 > c_2$

$f(\mathbf{x}) = c_2$

$f(\mathbf{x}) = c_1$

$f(\mathbf{x}) = c_0$

**Figure 2.15**  Slow convergence of steepest descent due to elongated cost contours (adapted from [Bertsekas, 1995]).

2. $\mathbf{D}^k = (\nabla^2 f(\mathbf{x}^k))^{-1}$ (*Newton method*): Recall that $\nabla^2 f(\mathbf{x}^k)$ is positive semidefinite for convex $f$. Suppose that for now that $\nabla^2 f(\mathbf{x}^k)$ is positive definite. The idea behind the Newton method is to minimize at iteration $k$ the quadratic approximation of $f$ around $\mathbf{x}^k$ as given below.

$$f^k(\mathbf{x}) = f(\mathbf{x}^k) + \nabla f(\mathbf{x}^k)^{\mathrm{T}}(\mathbf{x} - \mathbf{x}^k) + \frac{1}{2}(\mathbf{x} - \mathbf{x}^k)^{\mathrm{T}} \nabla^2 f(\mathbf{x}^k)(\mathbf{x} - \mathbf{x}^k).$$

Note that $\nabla f^k(\mathbf{x}) = \nabla f(\mathbf{x}^k) + \nabla^2 f(\mathbf{x}^k)(\mathbf{x} - \mathbf{x}^k)$. Solving $\nabla f^k(\mathbf{x}) = \mathbf{0}$ yields $\mathbf{x} = \mathbf{x}^k - (\nabla^2 f(\mathbf{x}^k))^{-1} \nabla f(\mathbf{x}^k)$, which is generalized in the Newton method as

$$\mathbf{x}^{k+1} = \mathbf{x}^k - \alpha^k (\nabla^2 f(\mathbf{x}^k))^{-1} \nabla f(\mathbf{x}^k). \tag{2.25}$$

Figure 2.16 illustrates the Newton method with $\alpha^k = 1$ for all $k$ (referred to as the *pure Newton method*).

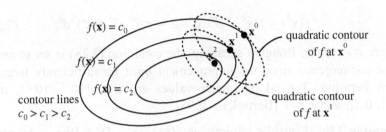

**Figure 2.16** Illustration of the pure Newton method.

When $\nabla^2 f(x^k)$ is positive semidefinite, it may not be invertible. In this case, a Newton step can be replaced by a steepest descent step [Bertsekas, 1995].

## Stepsize Selection

Four common choices for $\alpha^k$ are given below. Other choices of stepsizes can be found in [Bertsekas, 1995].

1. *Constant stepsize*: The simplest choice is to set $\alpha^k = s$ for some $s > 0$. The value of $s$ must be carefully chosen. If $s$ is too large, then the numerical method may not converge. On the other hand, if $s$ is too small, the convergence may be too slow. Therefore, a constant stepsize can be used only when it is not difficult to obtain a good value of $s$, e.g. by trials and errors.

2. *Minimization rule*: Given $d^k$, $\alpha^k$ is selected as

$$\alpha^k = \arg\min_{\alpha \geq 0} f(x^k + \alpha d^k). \qquad (2.26)$$

3. *Limited minimization rule*: For simpler implementation, the minimization rule can be modified to limit $\alpha^k$ to be in the interval $(0, s]$ for some $s > 0$, i.e.

$$\alpha^k = \arg\min_{\alpha \in (0, s]} f(x^k + \alpha d^k). \qquad (2.27)$$

4. *Successive stepsize reduction* or *Armijo rule*: To avoid the one-dimensional minimization required in the minimization and

limited minimization rules, the Armijo rule iteratively reduces the stepsize from $s > 0$ by a factor $\beta \in (0, 1)$ until the resultant stepsize, say $\beta^m s$, satisfies

$$f(\mathbf{x}^k) - f(\mathbf{x}^k + \beta^m s \mathbf{d}^k) \geq - \sigma \beta^m \, s \nabla \, f(\mathbf{x}^k)^\mathrm{T} \mathbf{d}^k, \qquad (2.28)$$

where $\sigma \in (0, 1)$. Roughly speaking, the condition (2.28) is set to avoid slow convergence since the improvement must be sufficiently large in each iteration. Typical parameter values are $\sigma \in [10^{-5}, 10^{-1}]$, $\beta \in [0.1, 0.5]$, and $s = 1$ [Bertsekas, 1995].

**Example 2.9:**   Consider minimizing $f(\mathbf{x}) = (x_1 - 4)^2 + 10(x_2 - 4)^2$ using numerical methods with the starting point $\mathbf{x}^0 = (0, 0)$. Notice that the cost function is quadratic with the minimum at $\mathbf{x}^* = (4, 4)$. Each numerical method is iterated until the marginal improvement is below 0.001. Figure 2.17 illustrates the iterations taken by the steepest descent method with a constant stepsize $s$. Notice how a large stepsize can cause a convergence problem.

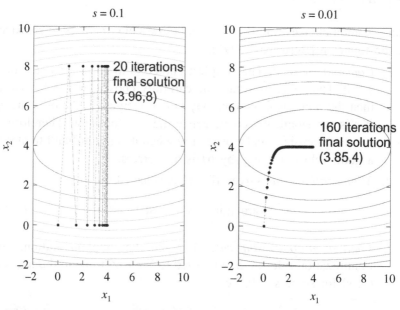

**Figure 2.17**    Iterations of the steepest descent method with a constant stepsize.

Figure 2.18 illustrates the iterations taken by the Newton method with a constant stepsize. Notice how the descent direction is towards the center of the quadratic cost function. In particular, when $s = 1$, the Newton method converges in a single step for the quadratic cost function.

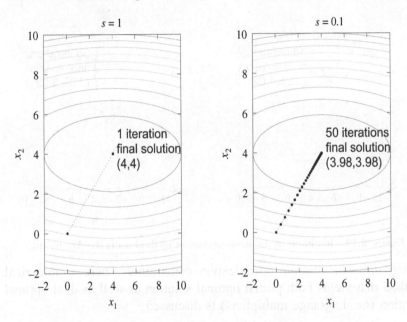

**Figure 2.18** Iterations of the Newton method with a constant stepsize.

Figure 2.19 illustrates the iterations taken by the steepest descent method using the Armijo rule for stepsize selection. Notice how the stepsize decreases as the method converges to the optimal solution. □

An important issue for numerical algorithms is the convergence analysis, which provides conditions under which the algorithms are guaranteed to converge. Such analysis is beyond the scope of this book and can be found in [Bertsekas, 1995; Boyd and Vandenberghe, 2004].

## 2.10 Numerical Algorithms for Constrained Optimization

This section discusses numerical algorithms for constrained optimization problems. The previously discussed gradient method is extended by

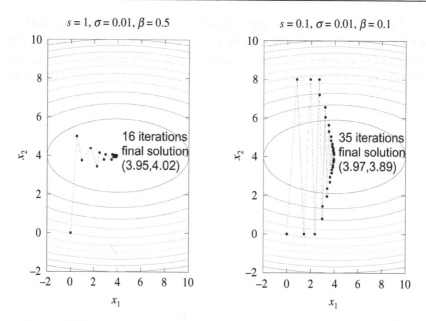

**Figure 2.19**  Iterations of the steepest descent method using the Armijo rule.

taking into account the optimization constraints. Then, a numerical method that yields both primal optimal solution as well as dual optimal solution (i.e. Lagrange multipliers) is discussed.

## Feasible Directions for Gradient Methods

Consider the convex optimization problem in (2.7). To apply the iterative descent approach, given a current solution $\mathbf{x}^k$, the next solution $\mathbf{x}^{k+1}$ is chosen such that (*i*) $f(\mathbf{x}^{k+1}) < f(\mathbf{x}^k)$, and (*ii*) $\mathbf{x}^{k+1} \in \mathcal{F}$, where $\mathcal{F}$ denotes the feasible set. Note that the second condition is added to what was previously considered for unconstrained optimization.

Given a feasible direction $\mathbf{d}^k$, a stepsize can be chosen using one of the methods previously discussed, i.e. constant stepsize, minimization rule, limited minimization rule, and Armijo rule. In what follows, methods for choosing $\mathbf{d}^k$ will be discussed.

Motivated by the previous consideration based on the first-order Taylor series approximation that $\nabla f(\mathbf{x}^k)^{\mathsf{T}} \mathbf{d}^k < 0$ is desirable, the *conditional gradient method* or the *Frank-Wolfe method* sets $\mathbf{d}^k = \tilde{\mathbf{x}}^k - \mathbf{x}^k$, where $\tilde{\mathbf{x}}^k$ is an optimal solution to the following optimization problem.

$$\text{minimize } \nabla f(\mathbf{x}^k)^T (\mathbf{x} - \mathbf{x}^k)$$

$$\text{subject to } \mathbf{x} \in \mathcal{F} \tag{2.29}$$

Given the stepsize $\alpha^k$, the next solution point is $\mathbf{x}^{k+1} = \mathbf{x}^k + \alpha^k (\tilde{\mathbf{x}}^k - \mathbf{x}^k)$. However, for the method to be practical, the problem must be such that subproblem (2.29) is relatively easy to solve.

The *gradient projection method* is based on projecting a vector on the feasible set. In particular, from the current solution $\mathbf{x}^k$, a feasible direction $\mathbf{d}^k$ is set to $\mathbf{d}^k = \tilde{\mathbf{x}}^k - \mathbf{x}^k$,

where

$$\tilde{\mathbf{x}}^k = (\mathbf{x}^k - \gamma^k \nabla f(\mathbf{x}^k))_{|\mathcal{F}} \tag{2.30}$$

for some $\gamma^k > 0$.

**Example 2.10:**   Consider the problem of minimizing $f(\mathbf{x}) = \dfrac{1}{2}(x_1^2 + x_2^2)$ in the set $\mathcal{F}$ as shown in figure 2.20. Assume that $\mathbf{x}^0 = (0, 2)$. Note that $\nabla f(\mathbf{x}^0) = (0, 2)$.

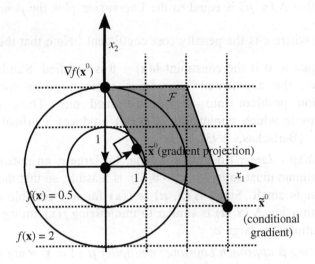

**Figure 2.20**   Finding feasible directions under the conditional gradient method and the gradient projection method.

Under the conditional gradient method, the subproblem to minimize $2(x_2 - 2)$ subject to $\mathbf{x} \in \mathcal{F}$ is considered. In this example, the subproblem can be solved by inspection to get $\tilde{\mathbf{x}}^0 = (3, -1)$, which is the point in $\mathcal{F}$ with the minimum value of $x_2$.

Under the gradient projection method with $\gamma^0 = 1$, the feasible direction is obtained by projecting $\mathbf{x}^0 - \nabla f(\mathbf{x}^0) = (0, 0)$ on $\mathcal{F}$, yielding $\tilde{\mathbf{x}}^0 = (0.8, 0.4)$.  □

## Augmented Lagrangian Method for Constrained Optimization

The numerical methods discussed so far only produce primal optimal solutions. This section discusses a method that yields both primal and dual optimal solutions. Consider for now the convex optimization problem in (2.7) but with only equality constraints. Define the *augmented Lagrangian* function, denoted by $\Lambda_c(\mathbf{x}, \boldsymbol{\mu})$, as follows.

$$\Lambda_c(\mathbf{x}, \boldsymbol{\mu}) = f(\mathbf{x}) + \boldsymbol{\mu}^\mathrm{T} \mathbf{h}(\mathbf{x}) + \frac{c}{2} \|\mathbf{h}(\mathbf{x})\|^2 \qquad (2.31)$$

Note that $\Lambda_c(\mathbf{x}, \boldsymbol{\mu})$ is equal to the Lagrangian plus the *penalty term* $\frac{c}{2}\|\mathbf{h}(\mathbf{x})\|^2$, where $c$ is the penalty cost coefficient.[3] Note that the penalty term is equal to 0 if the constraint $\mathbf{h}(\mathbf{x}) = \mathbf{0}$ is satisfied. Similar to the Lagrangian, the augmented Lagrangian turns a given constrained optimization problem into an unconstrained one. There are two mechanisms in which minimizing $\Lambda_c(\mathbf{x}, \boldsymbol{\mu})$ produces a primal optimal solution $\mathbf{x}^*$ [Bertsekas, 1995].

1. *Making c large*: Roughly speaking, for large $c$, an unconstrained minimum must satisfy $\mathbf{h}(\mathbf{x}) \approx \mathbf{0}$, i.e. is feasible, so that the penalty term is small. Since $\Lambda_c(\mathbf{x}, \boldsymbol{\mu}) = f(\mathbf{x})$ for any feasible solution, minimizing $\Lambda_c(\mathbf{x}, \boldsymbol{\mu})$ is similar to minimizing $f(\mathbf{x})$ among feasible solutions for large $c$.

2. *Making $\boldsymbol{\mu}$ approach Lagrange multiplier $\boldsymbol{\mu}^*$*: Let $\mathbf{x}' = \arg \min_{\mathbf{x} \in \mathbb{R}^N} \Lambda_c(\mathbf{x}, \boldsymbol{\mu}^*)$, it is argued below that

$$\Lambda_c(\mathbf{x}^*, \boldsymbol{\mu}^*) = \Lambda(\mathbf{x}^*, \boldsymbol{\mu}^*) \le \Lambda(\mathbf{x}', \boldsymbol{\mu}^*) \le \Lambda_c(\mathbf{x}', \boldsymbol{\mu}^*),$$

---

[3]In general, the augmented Lagrangian may contain a different form of the penalty term. However, this quadratic penalty is often used since it is mathematically tractable.

which implies that a primal optimal solution $\mathbf{x}^*$ is also a minimum of $\Lambda_c(\mathbf{x}, \mu^*)$. The first equality follows from the fact that $\Lambda_c(\mathbf{x}, \mu^*)$ $= \Lambda(\mathbf{x}, \mu^*)$ for any feasible solution $\mathbf{x}$, and hence also for $\mathbf{x}^*$. The first inequality follows from the Lagrangian optimality, which states that $\Lambda(\mathbf{x}^*, \mu^*) \leq \Lambda(\mathbf{x}, \mu^*)$ for any $\mathbf{x} \in \mathbb{R}^N$. The last inequality follows from the fact that the penalty term is nonnegative, and hence $\Lambda(\mathbf{x}, \mu^*) \leq \Lambda_c(\mathbf{x}, \mu^*)$ for any $\mathbf{x} \in \mathbb{R}^N$.

**Example 2.11 [Bertsekas, 1995]:**   Let $\mathbf{x} \in \mathbb{R}^2$. Consider minimizing $f(\mathbf{x}) = \dfrac{1}{2}(x_1^2 + x_2^2)$ subject to $x_1 = 1$. Using the KKT conditions, it is straightforward to verify that the primal-dual optimal solution pair are $\mathbf{x}^* = (1, 0)$ and $\mu^* = -1$. The augmented Lagragian is

$$\Lambda_c(x_1, x_2, \mu) = \frac{1}{2}(x_1^2 + x_2^2) + \mu(x_1 - 1) + \frac{c}{2}(x_1 - 1)^2.$$

By solving $\nabla\Lambda_c(x_1, x_2, \mu) = 0$, the unconstrained minimum is found to be

$$\tilde{x}_1(\mu, c) = \frac{c - \mu}{c + 1}, \quad \tilde{x}_2(\mu, c) = 0. \tag{2.32}$$

From (2.32), $c \to \infty$ makes $(\tilde{x}_1, \tilde{x}_2) \to (1, 0) = \mathbf{x}^*$. Thus, with $c \to \infty$, the unconstrained minimum approaches the constrained minimum, as illustrated in figure 2.21.

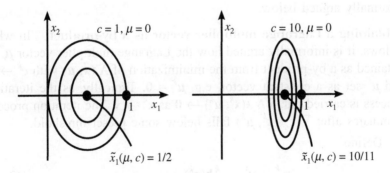

**Figure 2.21**   Convergence of the unconstrained minimum to the constrained minimum as $c \to \infty$.

In addition, from (2.32), $\mu \to -1$ makes $(\tilde{x}_1, \tilde{x}_2) \to (1, 0) = \mathbf{x}^*$. Thus, with $\mu \to \mu^*$, the unconstrained minimum approaches the constrained minimum, as illustrated in figure 2.22. □

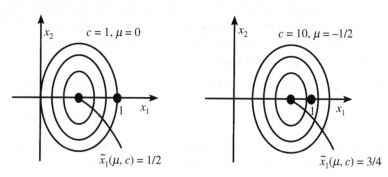

**Figure 2.22**  Convergence of the unconstrained minimum to the constrained minimum as $\mu \to \mu^*$.

Early numerical methods rely on increasing $c$ in each iteration and finding the limit point to which the sequence of solutions converges. In these methods, $\mu$ can simply be set to the zero vector throughout the process. The solution in each iteration can be found using a method such as the Newton method. This usually requires setting $c^k$ to be quite large, possibly causing some computational difficulty in the iterative procedure, e.g. causing elongated contour lines as shown in figure 2.15. Although there is no attempt to update $\mu$, once the iteration process terminates, the Lagrange multiplier vector $\mu^*$ is obtained as a by-product, as is informally argued below.

†**Obtaining a Lagrange multiplier vector as a by-product:**  In what follows, it is informally argued how the Lagrange multiplier vector $\mu^*$ is obtained as a by-product from the minimization of $\Lambda_c(\mathbf{x}, \mu)$ with $c^k \to \infty$ and $\mu^k$ set as a constant vector, e.g. $\mu^k = \mathbf{0}$. Typically, as the iteration process is carried out, $\nabla \Lambda_{c^k}(\mathbf{x}^k, \mu^k) \to \mathbf{0}$ as $\mathbf{x}^k \to \mathbf{x}$. The iteration process terminates after $\nabla \Lambda_{c^k}(\mathbf{x}^k, \mu^k)$ falls below some preset threshold.

Define

$$\tilde{\mu}^k = \mu^k + c^k \mathbf{h}(\mathbf{x}^k). \tag{2.33}$$

From the definition of $\Lambda_{c^k}(\mathbf{x}^k, \mu^k)$,

$$\nabla\Lambda_{ck}(\mathbf{x}^k, \boldsymbol{\mu}^k) = \nabla f(\mathbf{x}^k) + \nabla\mathbf{h}(\mathbf{x}^k)(\boldsymbol{\mu}^k + c^k\mathbf{h}(\mathbf{x}^k))$$

$$= \nabla f(\mathbf{x}^k) + \nabla\mathbf{h}(\mathbf{x}^k)\widetilde{\boldsymbol{\mu}}^k. \qquad (2.34)$$

Multiplying both sides of (2.34) by $(\nabla\mathbf{h}(\mathbf{x}^k)^T\nabla\mathbf{h}(\mathbf{x}^k))^{-1}\ \nabla\mathbf{h}(\mathbf{x}^k)^T$ (assuming that the inverse matrix exists) yields

$$\widetilde{\boldsymbol{\mu}}^k = (\nabla\mathbf{h}(\mathbf{x}^k)^T\nabla\mathbf{h}(\mathbf{x}^k))^{-1}\ \nabla\mathbf{h}(\mathbf{x}^k)^T(\nabla\Lambda_{ck}(\mathbf{x}^k, \boldsymbol{\mu}^k) - \nabla f(\mathbf{x}^k)). \quad (2.35)$$

Since $\nabla\Lambda_{ck}(\mathbf{x}^k, \boldsymbol{\mu}^k) \to \mathbf{0}$ as $\mathbf{x}^k \to \mathbf{x}^*$, (2.35) implies that $\widetilde{\boldsymbol{\mu}}^k \to \boldsymbol{\mu}^*$, where

$$\boldsymbol{\mu}^* = -(\nabla\mathbf{h}(\mathbf{x}^*)^T\nabla\mathbf{h}(\mathbf{x}^*))^{-1}\ \nabla\mathbf{h}(\mathbf{x}^*)^T\nabla f(\mathbf{x}^*).$$

From (2.34), $\nabla f(\mathbf{x}^*) + \nabla\mathbf{h}(\mathbf{x}^*)\boldsymbol{\mu}^* = \mathbf{0}$. In addition, since $\boldsymbol{\mu}^k$ is constant, the fact that $\widetilde{\boldsymbol{\mu}}^k = \boldsymbol{\mu}^k + c^k\mathbf{h}(\mathbf{x}^k)$ converges implies that $c^k\mathbf{h}(\mathbf{x}^k)$ converges. Since $c^k \to \infty$, it follows that $\mathbf{h}(\mathbf{x}^k) \to \mathbf{0}$, yielding $\mathbf{h}(\mathbf{x}^*) = \mathbf{0}$.

In conclusion, the KKT conditions are satisfied. Hence, $(\mathbf{x}^*, \boldsymbol{\mu}^*)$ is the primal-dual optimal solution pair with $\boldsymbol{\mu}^*$ being a Lagrange multiplier vector.

The convergence of $\boldsymbol{\mu}^k + c^k\mathbf{h}(\mathbf{x}^k)$ to $\boldsymbol{\mu}^*$ motivates the *method of multipliers* that updates both $c^k$ and $\boldsymbol{\mu}^k$ in each iteration.

## *Method of Multipliers (with Equality Constraints Only)*

Set the values of $\boldsymbol{\mu}^0$, $c^0$, and $\beta > 1$. Perform the following computations in iteration $k \in \mathbb{Z}^+$.

1. $\mathbf{x}^k = \arg\min_{\mathbf{x} \in \mathbb{R}^N} \Lambda_{ck}(\mathbf{x}, \boldsymbol{\mu}^k)$,
2. $c^{k+1} = \beta c^k$,
3. $\boldsymbol{\mu}^{k+1} = \boldsymbol{\mu}^k + c^k\ \mathbf{h}(\mathbf{x}^k)$

Appropriate choices of $\boldsymbol{\mu}^0$, $c^0$, and $\beta > 1$ vary from problem to problem. In practice, some trials and errors are typically required to obtain appropriate parameter values. More detailed discussions on how to select these parameters can be found in [Bertsekas, 1995].

**Example 2.12:** Let $\mathbf{x} \in \mathbb{R}^2$. Consider minimizing $f(\mathbf{x}) = \dfrac{1}{2}(x_1^2 + x_2^2)$ subject to $x_1 = 1$. Using the KKT conditions, it is straightforward to verify that the primal-dual optimal solution pair are $\mathbf{x}^* = (1, 0)$ and $\boldsymbol{\mu}^* = -1$. Below are some example operations of the multiplier method with $c^k = 2^k$ and $\boldsymbol{\mu}^0 = 0$.

The augmented Lagrangian is

$$\Lambda_c(\mathbf{x}, \mu) = \frac{1}{2}(x_1^2 + x_2^2) + \mu(x_1 - 1) + \frac{c}{2}(x_1 - 1)^2.$$

In step $k$, $\mathbf{x}^k$ and $\mu^k$ are computed as follows.

$$\mathbf{x}^k = \arg \min_{\mathbf{x} \in \mathbb{R}^2} \frac{1}{2}(x_1^2 + x_2^2) + \mu(x_1 - 1) + \frac{c}{2}(x_1 - 1)^2 = \left( \frac{c^k - \mu^k}{c^k + 1}, 0 \right)$$

$$\mu^{k+1} = \mu^k + c^k(x_1^k - 1)$$

Figure 2.23 shows the numerical values of $x_1^k$, $x_2^k$, and $\mu^k$ for different iterations. Observe that $\mathbf{x}^k \to (1, 0)$ and $\mu^k \to -1$ as expected.    □

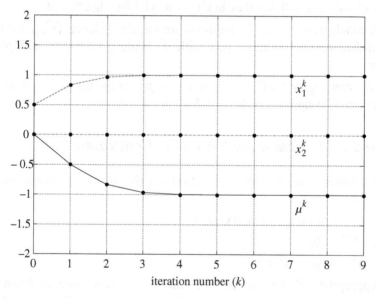

**Figure 2.23**  Numerical results from the multiplier method for minimizing
$$\frac{1}{2}(x_1^2 + x_2^2) \text{ subject to } x_1 = 1.$$

It remains to discuss the convex optimization problem in (2.7) with inequality constraints as well as equality constraints. The problem in (2.7) can be converted into one with equality constraints only by using additional variables $z_1$, ..., $z_L \in \mathbb{R}$ as follows.

First, the problem in (2.7) is rewritten as

$$\text{minimize } f(\mathbf{x})$$
$$\text{subject to } \forall l \in \{1, ..., L\}, \ g_l(\mathbf{x}) + z_l^2 = 0$$
$$\forall m \in \{1, ..., M\}, \ \mathbf{h}_m(\mathbf{x}) = 0$$

The augmented Lagrangian is therefore

$$\Lambda_c(\mathbf{x}, \mathbf{z}, \lambda, \mu) = f(\mathbf{x}) + \sum_{l=1}^{L} \left( \lambda_l (g_l(\mathbf{x}) + z_l^2) + \frac{c}{2} \left| g_l(\mathbf{x}) + z_l^2 \right|^2 \right)$$

$$+ \mu^{\mathrm{T}} \mathbf{h}(\mathbf{x}) + \frac{c}{2} \|\mathbf{h}(\mathbf{x})\|^2.$$

Observe that minimizing $\Lambda_c(\mathbf{x}, \mathbf{z}, \lambda, \mu)$ with respect to $\mathbf{z}$ can be done in closed form. Let $u_l = z_l^2$. Minimizing $\Lambda_c(\mathbf{x}, \mathbf{z}, \lambda, \mu)$ with respect to $\mathbf{z}$ yields

$$\min_{\mathbf{z} \in \mathbb{R}^L} \Lambda_c(\mathbf{x}, \mathbf{z}, \lambda, \mu) = f(\mathbf{x}) + \mu^{\mathrm{T}} \mathbf{h}(\mathbf{x}) + \frac{c}{2} \|\mathbf{h}(\mathbf{x})\|^2$$

$$+ \sum_{l=1}^{L} \min_{u_l \geq 0} \left( \lambda_l (g_l(\mathbf{x}) + u_l) + \frac{c}{2} \left| g_l(\mathbf{x}) + u_l \right|^2 \right).$$

Each minimization in the last term involves a quadratic function of $u_l$. The minimum is given by $u_l^* = \max(0, -\lambda_l/c - g_l(\mathbf{x}))$, where $-\lambda_l/c - g_l(\mathbf{x})$ is the unconstrained minimum without the constraint $u_l \geq 0$. Let $\Lambda_c(\mathbf{x}, \lambda, \mu) = \min_{\mathbf{z} \in \mathbb{R}^L} \Lambda_c(\mathbf{x}, \mathbf{z}, \lambda, \mu)$. It is straightforward to verify that

$$\Lambda_c(\mathbf{x}, \lambda, \mu) = f(\mathbf{x}) + \mu^{\mathrm{T}} \mathbf{h}(\mathbf{x}) + \frac{c}{2} \|\mathbf{h}(\mathbf{x})\|^2$$

$$+ \frac{1}{2c} \sum_{l=1}^{L} ([\max(0, \lambda_l + c g_l(\mathbf{x}))]^2 - \lambda_l^2). \quad (2.36)$$

Similar to the fact that $\mu^k + c^k \mathbf{h}(\mathbf{x}^k)$ converges to a Lagrange multiplier vector $\mu^*$, it can be shown that $\max(0, \lambda_l^k + c^k g_l(\mathbf{x}^k))$ converges to a Lagrange multiplier $\lambda_l^*$. The justification of this convergence is omitted here. The convergence gives rise to the multiplier method described below.

### *Method of Multipliers*

Set the values of $\mu^0$, $\lambda^0$, $c^0$, and $\beta > 1$. Perform the following computations in iteration $k \in \mathbb{Z}^+$.

1. $\mathbf{x}^k = \arg \min_{\mathbf{x} \in \mathbb{R}^N} \Lambda_{c^k}(\mathbf{x}, \lambda^k, \mu^k)$,
2. $c^{k+1} = \beta c^k$,
3. $\lambda_l^{k+1} = \max(0, \lambda_l^k + c^k g_l(\mathbf{x}^k))$ for $l \in \{1, ..., L\}$,
4. $\mu^{k+1} = \mu^k + c^k \mathbf{h}(\mathbf{x}^k)$.

**Example 2.13:**   Consider minimizing $f(x) = (x + 1)^2$ subject to $x \geq 0$. From example 2.3, the primal-dual optimal solution pair is $(x^*, \lambda^*) = (0, 2)$. Below are example operations of the multiplier method in which $c^k = 2^k$ and $\lambda^0 = 0$.

The augmented Lagrangian is

$$\Lambda_c(x, \lambda) = (x + 1)^2 + \frac{1}{2c}([\max(0, \lambda - cx)]^2 - \lambda^2).$$

The expression for $x^k$ is derived in what follows.

$$x^k = \arg \min_{x \in \mathbb{R}} \Lambda_{c^k}(x, \lambda^k)$$

$$= \arg \min_{x \in \mathbb{R}} (x + 1)^2 + \frac{1}{2c^k} [\max(0, \lambda^k - c^k x)]^2.$$

Note that the constant term $-\lambda^2$ is omitted above. Two separate cases are considered.

1. For $x \geq \lambda^k / c^k$,

$$\arg \min_{x \geq \lambda^k / c^k} \Lambda_{c^k}(x, \lambda^k) = \arg \min_{x \geq \lambda^k / c^k} (x + 1)^2 = \frac{\lambda^k}{c^k},$$

yielding

$$\min_{x \geq \lambda^k / c^k} \Lambda_{c^k}(x, \lambda^k) = \left(\frac{\lambda^k + c^k}{c^k}\right)^2.$$

2. For $x < \lambda^k / c^k$,

$$\arg \min_{x \le \lambda^k / c^k} \Lambda_{ck}(x, \lambda^k) = \arg \min_{x \le \lambda^k / c^k} (x + 1)^2$$

$$+ \frac{(\lambda^k + c^k x)^2}{2c^k} = \frac{\lambda^k - 2}{2 + c^k},$$

yielding

$$\min_{x \le \lambda^k / c^k} \Lambda_{ck}(x, \lambda^k) = \frac{2 + c^k}{c^k} \left( \frac{\lambda^k + c^k}{2 + c^k} \right)^2.$$

Since $\dfrac{2 + c^k}{c^k} \left( \dfrac{\lambda^k + c^k}{2 + c^k} \right)^2 < \left( \dfrac{\lambda^k + c^k}{c^k} \right)^2$, it follows that $x^k = \dfrac{\lambda^k - 2}{2 + c^k}$.

In conclusion, in iteration $k$, $x^k$ and $\lambda^{k+1}$ are updated as follows.

$$x^k = \frac{\lambda^k - 2}{2 + c^k} \quad \text{and} \quad \lambda^{k+1} = \max(0, \, \lambda^k - c^k x^k).$$

Figure 2.24 shows the numerical values of $x^k$ and $\lambda^k$ for different iterations. Observe that $x^k \to 0$ and $\lambda^k \to 2$ as expected. □

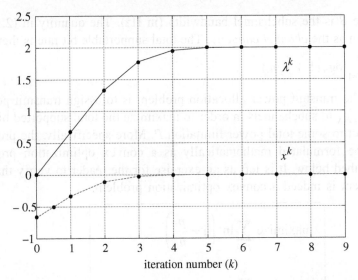

**Figure 2.24** Numerical results from the multiplier method for minimizing $(x + 1)^2$ subject to $x \ge 0$.

## 2.11    Application: Transmit Power Allocation

This section demonstrates an application of convex optimization to the problem of transmit power allocation in a multi-carrier communication system. Multi-carrier communication systems have become widespread, with notable examples being those using orthogonal frequency division multiplexing (OFDM). In a multi-carrier communication system, the bandwidth of a communication channel is divided into multiple narrow bands each of which is referred to as a *subchannel*. Since the bandwidth of each subchannel is narrow, the frequency response of a subchannel is approximately flat, leading to small signal distortion and thus small intersymbol interference (ISI).

Let $N$ be the number of subchannels in a multi-carrier system. Denote the noise levels on these subchannels by $n_1$, ..., $n_N$. Let $p_1$, ..., $p_N$ be the allocated transmit powers on the subchannels. Assuming the additive white Gaussian noise (AWGN) channel model for each subchannel, subchannel $i$ can support traffic at the bit rate of [Proakis and Salehi, 2008]

$$B \log_2 \left(1 + \frac{p_i}{n_i}\right) \text{ (in bps)}, \tag{2.37}$$

where $B$ is the subchannel bandwidth (in Hz). The quantity in (2.37) is known as the *channel capacity*. The total supportable bit rate is therefore

$$B \sum_{i=1}^{N} \log_2 (1 + p_i/n_i).$$

The transmit power allocation problem is to assign transmit powers $p_1$, ..., $p_N$ to subchannels in order to maximize the total supported bit rate subject to some total power limitation $P$. More specifically, the problem can be formulated mathematically as a convex optimization problem described below. It is left as an exercise for the reader to verify that the problem is indeed a convex optimization problem.

$$
\begin{aligned}
\text{maximize} \quad & \sum_{i=1}^{N} \ln \left(1 + \frac{p_i}{n_i}\right) \\
\text{subject to} \quad & \sum_{i=1}^{N} p_i \leq P \\
& \forall i \in \{1, ..., N\},\ p_i \geq 0
\end{aligned} \tag{2.38}
$$

Note that, in (2.38), the factor $B$ is omitted from the objective function. This is possible since a constant multiplication factor does not affect the optimality of a solution. In addition, the base-2 logarithmic function is replaced by the natural logarithmic function. This is possible since $\log_2 x = \ln x / \ln 2$, making the function change equivalent to omitting the constant multiplication factor of $1/\ln 2$ from the objective function.

For convenience, define $\mathbf{p} = (p_1, ..., p_N)$ and $\lambda = (\lambda_0, ..., \lambda_N)$. Viewing (2.38) as a maximization problem, the Lagrangian can be written as

$$\Lambda(\mathbf{p}, \lambda) = \sum_{i=1}^{N} \ln\left(1 + \frac{p_i}{n_i}\right) + \lambda_0 \left(P - \sum_{i=1}^{N} p_i\right) + \sum_{i=1}^{N} \lambda_i p_i.$$

From Lagrangian optimality in the KKT conditions, a primal-dual optimal solution pair $(\mathbf{p}^*, \lambda^*)$ must satisfy $\partial \Lambda(\mathbf{p}^*, \lambda^*)/\partial p_i = 0$, yielding

$$\frac{1}{p_i^* + n_i} - \lambda_0^* + \lambda_i^* = 0, \text{ or equivalently}$$

$$p_i^* + n_i = \frac{1}{\lambda_0^* - \lambda_i^*}.$$

The above condition together with complementary slackness, i.e. $\lambda_i^* p_i^* = 0$, yields

$$p_i^* > 0 \Rightarrow p_i^* + n_i = \frac{1}{\lambda_0^*}. \tag{2.39}$$

The condition in (2.39) states that the total signal-and-noise powers are equal among the subchannels to which nonzero transmit powers are allocated. In addition, since $p_i^* \geq 0$ and $n_i > 0$ in practice, (2.39) implies that $\lambda_0^* > 0$. From complimentary slackness, i.e. $\lambda_0^* \left(\sum_{i=1}^{N} p_i^* - P\right) = 0$, it follows that

$$\sum_{i=1}^{N} p_i^* = P, \tag{2.40}$$

which means that all the available transmit power is allocated.

The conditions in (2.39) and (2.40) yield an optimal power allocation process called *waterfilling*. The terminology is motivated by an analogy between allocating transmit powers to subchannels and pouring water into a container whose depths in $N$ different sections are specified by $n_1, ..., n_N$, as illustrated in figure 2.25.

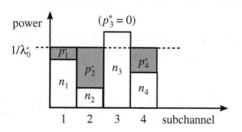

**Figure 2.25**    Transmit power allocation by waterfilling.

**Example 2.14:**    Suppose that $(n_1, n_2, n_3, n_4) = (3, 2, 1, 4)$ and $P = 10$. Then, the optimal transmit power allocation is given by $(p_1^*, p_2^*, p_3^*, p_4^*) = (2, 3, 4, 1)$. □

## Transmit Power Allocation for Frequency-selective Channels

Waterfilling can also be used to allocate transmit powers in a frequency-selective channel, as will be discussed next. Consider a frequency-selective channel in which the power gains on the $N$ subchannels are $h_1, ..., h_N$, where these gains are not necessarily equal. In addition, similar to before, let $n_1, ..., n_N$ be the noise levels on the $N$ subchannels.

Accordingly, the transmit power allocation problem is modified as follows.

$$
\begin{aligned}
&\text{maximize} \sum_{i=1}^{N} \ln\left(1 + \frac{h_i \, p_i}{n_i}\right) \\
&\text{subject to} \sum_{i=1}^{N} p_i \leq P \\
&\qquad \forall i \in \{1, ..., N\}, p_i \geq 0
\end{aligned}
$$

(2.41)

By a simple transformation, the problem in (2.41) can be made equivalent to the problem in (2.38). In particular, by defining the modified noise level $n_i' = n_i/h_i$, the problem can be expressed in terms of $p_1, ..., p_N$ and $n_1', ..., n_N'$, yielding the form in (2.38). Waterfilling can then be applied based on $n_1', ..., n_N'$ instead of $n_1, ..., n_N$.

**Example 2.15:** Consider an approximated frequency response of a communication channel shown in figure 2.26a. Suppose that $(n_1, ..., n_4) = (1, 1, 1, 1)$. Then, the problem in (2.41) can be converted to the problem in (2.38) by considering the modified noise levels as shown in figure 2.26b. More specifically, $(h_1, h_2, h_3, h_4) = (1/4, 1/8, 3/8, 1/2)$ and $(n_1, n_2, n_3, n_4) = (1, 1, 1, 1)$ are used to obtain $(n_1', n_2', n_3', n_4') = (4, 8, 8/3, 2)$.

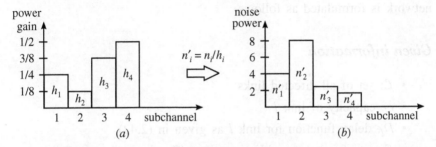

**Figure 2.26** Frequency-selective channel response and the problem transformation for waterfilling.

Suppose that $P = 20$. Then, waterfilling yields the optimal transmit power allocation $(p_1^*, p_2^*, p_3^*, p_4^*) = (31/6, 7/6, 39/6, 43/6)$. □

## 2.12 Application: Minimum Delay Routing

In this section, convex optimization is applied to solve the minimum delay routing problem in a packet-switched network. In a packet switched network, each node contains an electronic switch that can store and forward packets. Assume that all link capacities are known. In addition, the traffic demand (in bps) of each source-destination (s-d) pair is assumed known.

Let $\mathcal{L}$ be the set of directed links. Let $c_l$ be the capacity (in bps) of link $l$. Let $f_l$ be the total flow (summed over all s-d pairs) on link $l$. In general, the objective of routing is to minimize the function of the form

$$\sum_{l \in \mathcal{L}} D_l(f_l),$$

where $D_l$ is a monotonically nondecreasing function and serves effectively as the link cost. Below is the expression for $D_l$ which is obtained from queueing theory and reflects the average packet delay across link $l$ [Bertsekas and Gallager, 1992].

$$D_l(f_l) = \frac{f_l}{c_l - f_l} \qquad (2.42)$$

The problem of minimum delay routing in a packet-switched network is formulated as follows.

### Given information

- $\mathcal{L}$: set of all directed links
- $c_l$: capacity of link $l$
- $D_l$: delay function for link $l$ as given in (2.42)
- $S$: set of s-d pairs with nonzero traffic
- $t^s$: traffic demand (in packet/s) for s-d pair $s$
- $\mathcal{P}^s$: set of candidate paths for s-d pair $s$
- $\mathcal{P}_l$: set of candidate paths that use link $l$

### Variables

- $x^p$: traffic flow (in packet/s) on path $p$

### Constraints

- Satisfaction of traffic demands

$$\forall s \in \mathcal{S}, \ \sum_{p \in \mathcal{P}^s} x^p = t^s$$

- Non-negativity of traffic flows

$$\forall s \in \mathcal{S}, \forall p \in \mathcal{P}^s, x^p \geq 0$$

## *Objective*

- Minimize the overall packet delay in the network[4]

$$\text{minimize} \sum_{l \in \mathcal{L}} D_l \left( \sum_{p \in \mathcal{P}_l} x^p \right)$$

The overall optimization problem is as follows.

$$
\begin{array}{l}
\text{minimize} \sum_{l \in \mathcal{L}} D_l \left( \sum_{p \in \mathcal{P}_l} x^p \right) \\[2em]
\text{subject to } \forall s \in \mathcal{S}, \sum_{p \in \mathcal{P}^s} x^p = t^s \\[2em]
\forall s \in \mathcal{S}, \forall p \in \mathcal{P}^s, x^p \geq 0
\end{array}
\tag{2.43}
$$

Note that the formulation in (2.43) assumes sufficient link capacities in the network and does not contain explicit constraints for link capacities. Typically, the link delay $D_l$ approaches $\infty$ as the link utilization factor approaches 1. Therefore, any routing that leads to a link overflow will lead to an unusually high optimal cost and cannot be optimal.

## Characterization of Minimum Delay Routing

It can be verified that the function $D_l$ in (2.42) is convex and differentiable and that the minimum delay routing problem in (2.43) is a convex optimization problem. For convenience, let **x** be a vector that contains all flow values $x^p$. The Lagrangian can be written as

---

[4]Note that $f_l = \sum_{p \in \mathcal{P}_l} x^p$ in the above discussion.

$$\Lambda(\mathbf{x}, \boldsymbol{\lambda}, \boldsymbol{\mu}) = \sum_{l \in \mathcal{L}} D_l \left( \sum_{p \in \mathcal{P}_l} x^p \right) - \sum_{s \in \mathcal{S}} \sum_{p \in \mathcal{P}^s} \lambda^p x^p + \sum_{s \in \mathcal{S}} \mu^s \left( \sum_{p \in \mathcal{P}^s} x^p - t^s \right)$$

Applying Lagrangian optimality in the KKT conditions yields

$$\frac{\partial \Lambda(\mathbf{x}^*, \boldsymbol{\lambda}^*, \boldsymbol{\mu}^*)}{\partial x^p} = \sum_{l \in p} \frac{\partial D_l \left( \sum_{p \in \mathcal{P}_l} x^{p*} \right)}{\partial x^p} - \lambda^{p*} + \mu^{s(p)*} = 0, \quad (2.44)$$

where "$l \in p$" means that link $l$ belongs to path $p$, and $s(p)$ denotes the s-d pair corresponding to path $p$. For convenience, $D_l \left( \sum_{p \in \mathcal{P}_l} x^{p*} \right)$ will be denoted as $D_l(\mathbf{x}^*)$ in what follows.

From complimentary slackness in the KKT conditions, i.e. $\lambda^{p*} x^{p*} = 0$ for all $s \in \mathcal{S}$ and $p \in \mathcal{P}^s$, it follows from (2.44) that

$$x^{p*} > 0 \Rightarrow \sum_{l \in p} \frac{\partial D_l(\mathbf{x}^*)}{\partial x^p} = -\mu^{s(p)*}$$

$$x^{p*} = 0 \Rightarrow \sum_{l \in p} \frac{\partial D_l(\mathbf{x}^*)}{\partial x^p} = \lambda^{p*} - \mu^{s(p)*} \geq -\mu^{s(p)*} \quad (2.45)$$

One way to interpret (2.45) is to view $\sum_{l \in p} \partial D_l(\mathbf{x}^*)/\partial x^p$ as the length of path $p$ in which the length of link $l$ is equal to the first derivative $\partial D_l/\partial x^p$ evaluated at $\mathbf{x}^*$ [Bertsekas and Gallager, 1992]. Since $\mu^{s(p*)}$ is fixed for all paths of $s$, the condition in (2.45) states that all paths of $s$ with positive flows have the same first-derivative length. In addition, the length of each of these paths is equal to the *minimum first-derivative length (MFDL)*.

**Example 2.16 (Example 5.7 in [Bertsekas and Gallager, 1992]):** Consider the simple two-link network shown in figure 2.27. Assume that $c_1 > c_2$.

Let $r < c_1 + c_2$ be the traffic demand for the s-d pair. In addition, let $x_1$ and $x_2$ denote the traffic flows on links 1 and 2 respectively. Following

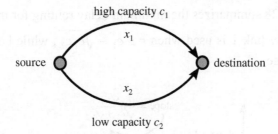

**Figure 2.27** Two-link network for minimum delay routing.

(2.42), $D_l(x_l) = \dfrac{x_l}{c_l - x_l}$ for $l \in \{1, 2\}$. It is useful to note that

$$\frac{\partial D_l(x_l)}{\partial x_l} = \frac{c_l}{(c_l - x_l)^2}.$$

There are two cases to consider: only link 1 carrying traffic and both links carrying traffic.[5]

1. $x_1^* = r$ and $x_2^* = 0$: The MFDL condition in (2.45) yields

$$\frac{\partial D_1(r)}{\partial x_1} \leq \frac{\partial D_2(0)}{\partial x_2} \Rightarrow \frac{c_1}{(c_1 - r)^2} \leq \frac{1}{c_2} \Rightarrow r \leq c_1 - \sqrt{c_1 c_2}.$$

2. $x_1^*, x_2^* > 0$: The MFDL condition in (2.45) yields

$$\frac{\partial D_1(x_1^*)}{\partial x_1} = \frac{\partial D_2(x_2^*)}{\partial x_2} \Rightarrow \frac{c_1}{(c_1 - x_1^*)^2} = \frac{c_2}{(c_2 - x_2^*)^2}.$$

The above equation together with $x_1^* + x_2^* + r$ yields

$$x_1^* = \frac{\sqrt{c_1}\,(r - (c_2 - \sqrt{c_1 c_2}))}{\sqrt{c_1} + \sqrt{c_2}},$$

$$x_2^* = \frac{\sqrt{c_2}\,(r - (c_1 - \sqrt{c_1 c_2}))}{\sqrt{c_1} + \sqrt{c_2}}.$$

---

[5]Since $c_1 > c_2$, it does not make sense to have only link 2 carry traffic.

Figure 2.28 summarizes the minimum delay routing for this example. Note that only link 1 is used when $r \leq c_1 - \sqrt{c_1 c_2}$, while both links are used otherwise.    □

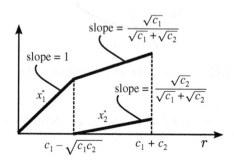

**Figure 2.28**    Minimum delay routing for the two-link network.

**Example 2.17:**    Consider the three-node network with link capacities shown in figure 2.29. Suppose there are two s-d pairs 1-3 and 2-3 whose paths are also shown in figure 2.29. Denote the traffic flows on these four paths by $x^1$, $x^2$, $x^3$, and $x^4$. Assume that the traffic demands for s-d pairs 1-3 and 2-3 are $t^1 = 2$ and $t^2 = 2$ respectively.

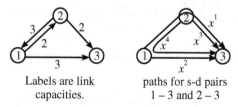

Labels are link                paths for s-d pairs
capacities.                    1 − 3 and 2 − 3

**Figure 2.29**    Three-node network for minimum delay routing.

Following (2.42), the objective function for minimum delay routing is

$$D_{(1,2)}(x^1) + D_{(2,1)}(x^4) + D_{(1,3)}(x^2 + x^4) + D_{(2,3)}(x^1 + x^3)$$

$$= \frac{x^1}{2 - x^1} + \frac{x^4}{3 - x^4} + \frac{x^2 + x^4}{3 - x^2 - x^4} + \frac{x^1 + x^3}{2 - x^1 - x^3}.$$

The method of multipliers will be applied to solve for primal and dual optimal solutions. The augmented Lagrangian is

$$\Lambda_c(\mathbf{x}, \lambda, \mu) = \frac{x^1}{2 - x^1} + \frac{x^4}{3 - x^4} + \frac{x^2 + x^4}{3 - x^2 - x^4} + \frac{x^1 + x^3}{2 - x^1 - x^3}$$

$$+ \mu^1(x^1 + x^2 - 2) + \mu^2(x^3 + x^4 - 2) + \frac{c}{2}(x^1 + x^2 - 2)^2$$

$$+ \frac{c}{2}(x^3 + x^4 - 2)^2 + \frac{1}{2c} \sum_{i=1}^{4} [(\max(0, \lambda^i - cx^i))^2 - (\lambda^i)^2].$$

To proceed with the method of multipliers, set $c^k = 2^k$ and initialize all Lagrange multipliers to zero. For unconstrained minimization of $\Lambda_c(\mathbf{x}, \lambda, \mu)$ in each iteration, the steepest descent method is used with a constant stepsize equal to 0.001. The iteration terminates when the marginal cost improvement is below 0.0001.

To use the steepest descent method, the following gradient is computed.

$$\nabla \Lambda = \begin{bmatrix} \partial \Lambda_c / \partial x^1 \\ \partial \Lambda_c / \partial x^2 \\ \partial \Lambda_c / \partial x^3 \\ \partial \Lambda_c / \partial x^4 \end{bmatrix}$$

$$= \begin{bmatrix} \frac{2}{(2 - x^1)^2} + \frac{2}{(2 - x^1 - x^3)^2} + \mu^1 + c(x^1 + x^2 - 2) - \max(0, \lambda^1 - cx^1) \\ \frac{3}{(3 - x^2 - x^4)^2} + \mu^1 + c(x^1 + x^2 - 2) - \max(0, \lambda^2 - cx^2) \\ \frac{2}{(2 - x^1 - x^3)^2} + \mu^2 + c(x^3 + x^4 - 2) - \max(0, \lambda^3 - cx^3) \\ \frac{3}{(3 - x^4)^2} + \frac{3}{(3 - x^2 - x^4)^2} + \mu^2 + c(x^3 + x^4 - 2) - \max(0, \lambda^4 - cx^4) \end{bmatrix}$$

Figure 2.30 shows the numerical results obtained from the method of multipliers. In particular, the optimal cost is 8.07. The primal and dual optimal solutions are

$$(x^{1*}, x^{2*}, x^{3*}, x^{4*}) = (0, 2, 1.55, 0.448),$$
$$(\lambda^{1*}, \lambda^{2*}, \lambda^{3*}, \lambda^{4*}, \mu^{1*}, \mu^{2*}) = (0.40, 0, 0, 0, -9.95, -10.1). \qquad \square$$

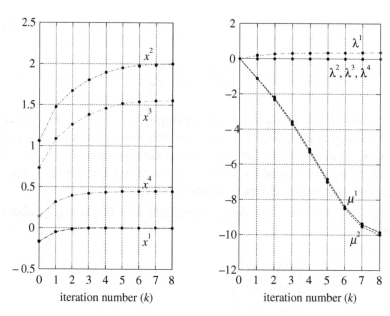

**Figure 2.30**   Numerical results for minimum delay routing from the method of multipliers.

## 2.13   Exercise Problems

**Problem 2.1 (Convexity of a set)**   In each case, identify whether a given set $\mathcal{X}$ is a convex set.

(a) $\mathcal{X} = \{\mathbf{x} \in \mathbb{R}^2 \mid x_1 \leq a \text{ or } x_2 \leq b\}$, where $a, b \in \mathbb{R}$

(b) $\mathcal{X} = \{\mathbf{x} \in \mathbb{R}^N \mid \mathbf{a}^T\mathbf{x} = b\}$, where $\mathbf{a} \in \mathbb{R}^N$ and $b \in \mathbb{R}$

   **NOTE:** The set of this form is called an $N$-dimensional hyperplane.

(c) $\mathcal{X} = \{\mathbf{x} \in \mathbb{R}^N \mid \mathbf{a}^T\mathbf{x} \geq b\}$, where $\mathbf{a} \in \mathbb{R}^N$ and $b \in \mathbb{R}$

   **NOTE:** The set of this form is called an $N$-dimensional halfspace.

(d) $\mathcal{X} = \{\mathbf{x} \in \mathbb{R}^N | \mathbf{Ax} \geq \mathbf{b}, \mathbf{Cx} = \mathbf{d}\}$, where $\mathbf{A}$ is an $L \times N$ matrix, $\mathbf{C}$ is an $M \times N$ matrix, $\mathbf{b} \in \mathbb{R}^L$, and $\mathbf{d} \in \mathbb{R}^M$

NOTE: The set of this form is called an $N$-dimensional polyhedron, which is a finite intersection of halfspaces and hyperplanes.

(e) $\mathcal{X} = \{\mathbf{x} \in \mathbb{R}^N | \|\mathbf{x} - \mathbf{a}\| \leq r\}$, where $\mathbf{a} \in \mathbb{R}^N$ and $r > 0$

NOTE: The set of this form is called an $N$-dimensional ball with radius $r$ centered at point $\mathbf{a}$.

(f) $\mathcal{X} = \left\{ \varphi_1 \mathbf{v}_1 + \cdots \varphi_M \mathbf{v}_M \left| \sum_{i=1}^{M} \varphi_i = 1 \text{ and } \varphi_i \geq 0 \text{ for all } i \right. \right\}$, where $\mathbf{v}_1, ..., \mathbf{v}_M \in \mathbb{R}^N$

NOTE: The set of this form is called the convex hull of points $\mathbf{v}_1, ..., \mathbf{v}_M$.

(g) $\mathcal{X} = \bigcap_{m=1}^{M} \mathcal{A}_m$, where each $\mathcal{A}_m$ is a convex subset of $\mathbb{R}^N$

(h) $\mathcal{X} = \bigcup_{m=1}^{M} \mathcal{A}_m$, where each $\mathcal{A}_m$ is a convex subset of $\mathbb{R}^N$

**Problem 2.2 (Convexity of a function)** Consider a real function $f$ defined over a convex set $\mathcal{X}$. In each case, identify whether the given function is convex.

(a) Let $\mathcal{X} = \mathbb{R}$. Define $f(x) = x^3$.

(b) Let $\mathcal{X} = \mathbb{R}^+$. Define $f(x) = x^3$.

(c) Let $\mathcal{X} = \mathbb{R}^2$. Define $f(\mathbf{x}) = \dfrac{1}{2}\mathbf{x}^T\mathbf{Ax} + \mathbf{b}^T\mathbf{x} + c$, where $\mathbf{A} = \begin{bmatrix} 1 & 3 \\ 3 & 5 \end{bmatrix}$,

$\mathbf{b} = \begin{bmatrix} 1 \\ 2 \end{bmatrix}$, and $c = -1$.

(d) Let $\mathcal{X} = \mathbb{R}$. Define $f(\mathbf{x}) = \dfrac{1}{2}\mathbf{x}^T\mathbf{Ax} + \mathbf{b}^T\mathbf{x} + c$, where $\mathbf{A} = \begin{bmatrix} 4 & 3 \\ 3 & 5 \end{bmatrix}$,

$\mathbf{b} = \begin{bmatrix} 1 \\ 2 \end{bmatrix}$, and $c = -1$.

(e) Let $\mathcal{X} = \mathbb{R}^N$. Define $f(\mathbf{x}) = \min(\mathbf{a}^T\mathbf{x}, \mathbf{b}^T\mathbf{x})$, where $\mathbf{a}, \mathbf{b} \in \mathbb{R}^N$.

(f) Let $\mathcal{X} = \mathbb{R}^N$. Define $f(\mathbf{x}) = \max(\mathbf{a}^T\mathbf{x}, \mathbf{b}^T\mathbf{x})$, where $\mathbf{a}, \mathbf{b} \in \mathbb{R}^N$.

**Problem 2.3 (Convexity of an optimization problem)**  Let $f$ and $g$ be real and convex functions defined over a convex set $\mathcal{X} \subset \mathbb{R}^N$. Show that the following optimization problem is a convex optimization problem.

$$\text{minimize } f(\mathbf{x})$$

$$\text{subject to } g(\mathbf{x}) \leq 0$$

**Problem 2.4 (Convexity of an optimization problem)**  Consider the optimization problem whose variables are $x_1, \ldots, x_N$ as shown below.

$$\text{minimize } \sum_{i=1}^{N} x_i \ln x_i$$

$$\text{subject to } \sum_{i=1}^{N} x_i = 1$$

$$x_i > 0, \ i \in \{1, \ldots, N\}$$

Show that this problem is a convex optimization problem.

**NOTE:**  The reader who is familiar with the entropy of a random variable may recall that the objective function is the negative of the entropy of a random variable whose probability mass function (PMF) is described by $x_1, \ldots, x_N$. Therefore, this problem is equivalent to finding the PMF that maximizes the entropy.

**Problem 2.5 (Using the KKT conditions)**  Let $s < 2$. Consider the following convex optimization problem whose decision variable is $x$.

$$\text{minimize } -xe^{-x}$$

$$\text{subject to } x \leq s$$

(a) Verify that the cost function is convex in the feasible set. (Hence, the problem can be viewed as a convex optimization problem.)

(b) Using the KKT conditions, find a primal-dual optimal solution pair for this problem.

**HINT:** The answer depends on the specific value of $s$.

**Problem 2.6 (Using the KKT conditions)**   Let $N$ be an integer with $N \geq 2$. Let $c_1,$ ..., $c_N$ be positive real numbers such that $\sum_{i=1}^{N} c_i = 1$. Consider the following optimization problem whose variables are $x_1,$ ..., $x_N$.

$$\text{minimize} \sum_{i=1}^{N} c_i x_i$$

$$\text{subject to} \sum_{i=1}^{N} e^{-x_i} \leq 1$$

(a) Show that the above problem is a convex optimization problem.

(b) Write down the dual function in terms of $N$, $c_1,$ ..., $c_N$, and the dual variables. In addition, write down the dual problem.

(c) Find a primal-dual optimal solution pair for this problem.

**NOTE:** The reader who is familiar with source coding may recall that the constraint is similar to the Kraft inequality for uniquely decodable codes, i.e. $\sum_{i=1}^{N} 2^{-x_i} = \sum_{i=1}^{N} e^{-x_i \ln 2} \leq 1$, where $x_1,$ ..., $x_N$ are codeword lengths. The values $c_1,$ ..., $c_N$ form the PMF of a source symbol. Hence, this problem is similar to finding the smallest average codeword length, but with $x_1,$ ..., $x_N$ allowed to be real values (not necessarily integers).

**Problem 2.7 (Convexification of an optimization problem)**   Let $a$ and $b$ be two positive integers. Consider the following optimization problem.

$$\text{minimize} - x_1^a x_2^b$$

$$\text{subject to } x_1 + x_2 \leq 1$$

$$x_1, x_2 \geq 0$$

(a) Is the above problem a convex optimization problem for any given positive integers $a$ and $b$?

(b) Consider the change of variables $y_1 = \ln x_1$ and $y_2 = \ln x_2$. Show that solving the optimization in part (a) is equivalent to solving the following problem.

$$\text{minimize} - ay_1 - by_2$$
$$\text{subject to } e^{y_1} + e^{y_2} \leq 1$$

(c) Show that the optimization problem in part (b) is a convex optimization problem.

(d) Find the dual function and the dual problem for the optimization problem in part (b).

(e) Find primal and dual optimal solutions for the optimization problem in part (b).

(f) Specify an optimal solution and the optimal cost for the original optimization problem in part (a).

**Problem 2.8 (Using the sensitivity information)**    Consider a plan to produce chairs and tables to sell at the prices of 3 and 4 dollars per unit respectively. In addition, there are two workers, say A and B. Making a chair requires 2 hours of work from A and 4 hours of work from B, while making a table requires 3 hours of work from A and 2 hours of work from B. Each worker works 10 hours a day. The optimization problem is to maximize the revenue per day by deciding on the number of chairs and tables to produce.

Let $x_1$ and $x_2$ denote the numbers of chairs and tables to be produced respectively. For simplicity, $x_1$ and $x_2$ can be nonintegers, i.e. unfinished works can be continued the next day. The optimization problem is given below.

$$\text{minimize } 3x_1 + 4x_2$$
$$\text{subject to } 2x_1 + 3x_2 \leq 10$$
$$4x_1 + 2x_2 \leq 10$$
$$x_1, x_2 \geq 0$$

(a) Find the primal-dual optimal solution pair for this problem.

(b) If either A or B can be hired to work an extra hour with the same amount of money, who should be hired based on the sensitivity information in the Lagrange multipliers? Compute the new primal optimal cost after his/her extra work is assigned based on the decision, and compare it to the new cost for the alternative decision.

**Problem 2.9 (Steepest descent and Newton methods)** Define $f(x_1, x_2) = x_1^4 + x_1^2 x_2^2 + x_2^4$.

(a) Show that $f$ is a convex function.

(b) Consider the steepest descent method for unconstrained minimization of $f$. Suppose that the stepsize is 0.1, while the inital point is $\mathbf{x}^0 = (1, 1)$. Compute the next two points $\mathbf{x}^1$ and $\mathbf{x}^2$ under this method. NOTE: Use a calculator to compute numerical values.

(c) Consider the Newton method for unconstrained minimization of $f$. Suppose that the stepsize is 0.1, while the initial point is $\mathbf{x}^0 = (1, 1)$. Compute the next two points $\mathbf{x}^1$ and $\mathbf{x}^2$ under this method.

**Problem 2.10 (Conditional gradient and gradient projection methods)** Define $f(x_1, x_2) = x_1^2 - x_1 x_2 + x_2^2$.

(a) Show that $f$ is a convex function.

(b) Consider the steepest descent method for unconstrained optimization of $f$. Suppose that the stepsize is 0.1, while the initial point is $\mathbf{x}^0 = (1, 2)$. Compute the next point $\mathbf{x}^1$ under this method. Consider now the constained optimization problem below.

$$\text{minimize } x_1^2 - x_1 x_2 + x_2^2$$
$$\text{subject to } x_1 - x_2 \leq 0$$
$$x_1 \geq -1$$

(c) Consider the conditional gradient method for constrained minimization of $f$. Suppose that the stepsize is 0.1, while the initial point is $\mathbf{x}^0 = (1, 2)$. Compute the next point $\mathbf{x}^1$ under this method. HINT: Draw the feasible set to help identify $\mathbf{x}^1$.

(d) Consider the gradient projection method for constrained minimization of $f$. Suppose that the stepsize before projection is 1, the stepsize after projection is 0.1, and the initial point is $\mathbf{x}^0 = (1, 2)$. Compute the next point $\mathbf{x}^1$ under this method. HINT: Draw the feasible set to help identify $\mathbf{x}^1$.

**Problem 2.11 (Waterfilling)**  Consider a multicarrier communication channel that contains 4 subchannels. Suppose that channel gains (power gains) are given by $(h_1, h_2, h_3, h_4) = (0.25, 0.5, 0.2, 0.5)$. In addition, suppose the noise powers for all subchannels are equal to 1 unit. The power allocation problem is to solve the following optimization problem whose variables are $p_1$, $p_2$, $p_3$, and $p_4$.

$$\text{maximize} \sum_{i=1}^{4} \log \left(1 + \frac{h_i p_i}{n_i}\right)$$

$$\text{subject to} \sum_{i=1}^{4} p_i \leq P$$

$$p_1, p_2, p_3, p_4 \geq 0$$

(a) Let $P = 10$. Find a primal-dual optimal solution pair for the given problem.

(b) Repeat part (a) for $P = 5$.

**Problem 2.12 (Power allocation with multiple users)**  Consider the problem of multi-user power allocation described as follows. There are $I$ transmission subchannels to be shared by $J$ users. Each subchannel can be assigned to at most one user. Let $n_i$ denote the noise power on subchannel $i$. Define the following decision variables.

- $x_i^j \in \{0, 1\}$: equal to 1 if and only if subchannel $i$ is assigned to user $j$
- $p_i^j \geq 0$: power level assigned to subchannel $i$ and to user $j$

The overall problem is as follows.

$$\text{maximize} \sum_{j=1}^{J} \sum_{i=1}^{I} \ln \left(1 + \frac{p_i^j}{n_i}\right)$$

$$\text{subject to} \sum_{j=1}^{J} \sum_{i=1}^{I} p_i^j \leq P$$

$$\forall i \in \{1, ..., I\}, \sum_{j=1}^{J} x_i^j \leq 1$$

$$\forall i \in \{1, ..., I\}, \forall j \in \{1, ..., J\}, p_i^j \le Px_i^j$$

$$\forall i \in \{1, ..., I\}, \forall j \in \{1, ..., J\}, x_i^j \in \{0, 1\}, p_i^j \ge 0$$

(a) Is the given problem a convex optimization problem?

(b) Count the number of variables and the number of constraints (not including the integer contraints of $x_i^j$ and the nonnegativity constraints of $p_i^j$) in terms of $I$ and $J$.

(c) Find one optimal power allocation (i.e. the set of values for $x_i^j$ and $p_i^j$) for the specific case with $J = 2$, $I = 4$, $P = 12$, and $(n_1, n_2, n_3, n_4) = (3, 1, 2, 2)$. HINT: Does it matter to whom each subchannel is assigned? Does the problem have a unique optimal answer?

(d) Suppose now that user $i$ experiences power gain $h_i^j$ on subchannel $j$, yielding the modified objective function

$$\text{maximize} \sum_{j=1}^{J} \sum_{i=1}^{I} \ln \left( 1 + \frac{h_i^j \, p_i^j}{n_i} \right)$$

Argue that, in any optimal power allocation, each subchannel $i$ is assigned to the user with the maximum gain $h_i^j$. HINT: It is possible to proceed by contradiction.

(e) For the modified objective function in part (d), find one optimal power allocation for the specific case with $J = 2$, $I = 4$, $P = 12$, $(n_1, n_2, n_3, n_4) = (3, 1, 2, 2)$, $(h_1^1, h_2^1, h_3^1, h_4^1) = (3, 1, 1, 1)$, and $(h_1^2, h_2^2, h_3^2, h_4^2) = (1, 1, 2, 1)$. HINT: Use the observation made in part (d).

(f) Suppose the new objective is to maximize the minimum data rate allocated to a user. Write down the problem objective modified from part (d). In addition, under this new objective, construct an example in which it is no longer optimal to assign each subchannel $i$ to the user with the maximum gain $h_i^j$.

**Problem 2.13 (Minimum delay routing)** Consider a three-link network shown in figure 2.31. Let the link capacities be $2c$, $c$, and $c$ respectively.

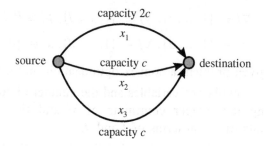

**Figure 2.31**    Three-link network for minimum delay routing.

Let $t < 4c$. Consider the following routing problem whose variables are the flows $x_1$, $x_2$, $x_3$ on the three links respectively.

$$\text{minimize} \quad \frac{x_1}{2c - x_1} + \frac{x_2}{c - x_2} + \frac{x_3}{c - x_3}$$

$$\text{subject to } x_1 + x_2 + x_3 = t$$

$$x_1, x_2, x_3 \geq 0$$

Find an optimal solution for the above routing problem. HINT: Your answer will depend on the specific value of $t$.

**Problem 2.14 (Network utility maximization)**    Consider a scenario in which users 1 and 2 share a transmission link whose capacity is $c$. Let $x_1$ and $x_2$ be the transmission rates assigned to users 1 and 2 respectively. Given $x_1$ and $x_2$, assume that the utilities or the happiness levels of users 1 and 2 are given by the utility functions $u_1(x_1)$ and $u_2(x_2)$ respectively. The utility maximization problem is to find a pair $(x_1, x_2)$ that is an optimal solution to the following optimization problem.

$$\text{maximize} \quad u_1(x_1) + u_2(x_2)$$

$$\text{subject to } x_1 + x_2 \leq c$$

$$x_1, x_2 \geq 0$$

(a) Assume that $u_1(x_1) = 1 - e^{-x_1}$ and $u_2(x_2) = 1 - e^{-x_2}$. Show that the objective function is a convex function.

(b) Assume the utility functions in part (a) together with $c = \ln 2$. Use the KKT conditions to find optimal primal and dual solutions for the above utility maximization problem.

(c) Assume that $u_1(x_1) = 1 - e^{-x_1}$ and $u_2(x_2) = 1 - e^{-x_2}$. together with $c = \ln 2$. Use the KKT conditions to find optimal primal and dual solutions for the above utility maximization problem.

(d) For the problem in part (c), find the value of $c$ below which the optimal primal solution will be $(x_1^*, x_2^*) = (0, c)$.

†**Problem 2.15 (Capacity of a DMC)**   Consider the problem of finding the capacity of a discrete memoryless channel (DMC). Let $f(x)$ and $f(y)$ denote the PMFs of the input and the output respectively. In addition, let $f(y|x)$ denote the conditional PMF of output $y$ given input $x$. Note that the DMC is fully described by $f(y|x)$.

Finding the capacity of a DMC is equivalent to solving for the optimal cost in the following convex optimization problem in which $f(y|x)$'s are given while $f(x)$'s are the variables. (It is straightforward to verify that the problem is convex.)

$$\text{maximize} \sum_x \sum_y f(y|x) f(x) \ln \frac{f(y|x)}{f(x)}$$

$$= \sum_x \sum_y f(y|x) f(x) \ln \frac{f(y|x)}{\sum_{x'} f(y|x') f(x')}$$

$$\text{subject to} \sum_x f(x) = 1$$

$$\forall x, f(x) \geq 0$$

(a) Let $C$ be the DMC capacity. Using the KKT conditions, show that the necessary and sufficient condition for $f(x)$ to be the capacity achieving PMF is to have

$$I(x) = C \text{ for all } f(x) > 0,$$

$$I(x) \leq C \text{ for all } f(x) = 0,$$

where $I(x) = \sum_y f(y|x) \ln \frac{f(y|x)}{f(y)}$. These KKT conditions can be used to verify that a given PMF achieves the DMC capacity.

(b) Use the above KKT conditions to compute numerically the capacity of the DMC shown in figure 2.32. Assume that $\varepsilon = 0.1$.

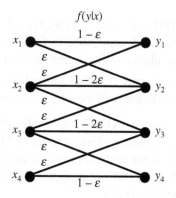

**Figure 2.32**   DMC for problem 2.15.

**Problem 2.16 (True or false)**   For each of the following statements, state whether it is true or false, i.e. not always true. If true, provide a brief justification. Otherwise, provide a counter-example.

(*a*) In a convex optimization problem, if there are two different feasible solutions, then there are infinitely many feasible solutions.

(*b*) In a convex optimization problem, if there are two different optimal solutions, then there are infinitely many optimal solutions.

(*c*) Consider a convex optimization problem whose objective is to minimize $f(x)$ subject to $x \in \mathcal{X}$. If $f$ is bounded in $\mathcal{X}$, then an optimal solution always exists.

(*d*) Let $f$ and $g$ be two real functions defined over a convex set. Suppose that both $f$ and $g$ are convex. Then, the function $h$ defined by $h(x) = f(x) \, g(x)$ is a convex function.

(*e*) Suppose $\mathbf{x}_1$ and $\mathbf{x}_2$ are two optimal solutions for the same convex optimization problem. Then, $\mathbf{x}_1 - \mathbf{x}_2$ is a feasible solution that yields the cost equal to zero.

# Chapter 3

# Linear Optimization

This chapter discusses optimization problems in which the objective and constraint functions are linear functions of decision variables. These problems have specific geometrical properties that can be exploited to construct a practically efficient algorithm called the simplex algorithm. Similar to convex optimization, duality is also discussed. While strong duality may not hold for convex optimization, it always holds for linear optimization when a primal optimal solution exists. Finally, specific applications of linear optimization are given, including minimum cost routing in wireline networks and energy efficient routing in wireless sensor networks.

## 3.1 Illustrative Example

The *general form* of a *linear optimization problem* is

$$\text{minimize } c_1 x_1 + \cdots + c_N x_N$$
$$\text{subject to } a_{11} x_1 + \cdots + a_{1N} x_N \geq b_1$$
$$\vdots$$
$$a_{M1} x_1 + \cdots + a_{MN} x_N \geq b_M$$

where $x_1, \ldots, x_N$ are the decision variables, $c_1, \ldots, c_N$ are the objective or cost coefficients, while $b_1, \ldots, b_M$ and $a_{11}, \ldots, a_{1N}, a_{21}, \ldots, a_{MN}$ are the constraint coefficients.

A *linear function f* of **x** has the form $f(\mathbf{x}) = \sum_{i=1}^{N} \alpha_i x_i$, where $\alpha_1$, ..., $\alpha_N$ are real coefficients. An *affine* function $f$ of **x** has the form $f(\mathbf{x}) = \alpha_0 + \sum_{i=1}^{N} \alpha_i x_i$, where $\alpha_0$ is the constant term. In the literature, the term "linear" is often used to include "affine" as well; this usage will be adopted in this book.

Note that, in the above linear optimization problem, the objective and constraint functions are linear combinations of the decision variables. For convenience, define the following matrix and vector notations.

$$\mathbf{x} = \begin{bmatrix} x_1 \\ \vdots \\ x_N \end{bmatrix}, \; \mathbf{c} = \begin{bmatrix} c_1 \\ \vdots \\ c_N \end{bmatrix}, \; \mathbf{b} = \begin{bmatrix} b_1 \\ \vdots \\ b_M \end{bmatrix}, \; \mathbf{A} = \begin{bmatrix} a_{11} & \cdots & a_{1N} \\ \vdots & \ddots & \vdots \\ a_{M1} & \cdots & a_{MN} \end{bmatrix} \quad (3.1)$$

Accordingly, the general form of a linear optimization problem can be expressed more compactly as shown below.[1]

$$\boxed{\begin{array}{l} \text{minimize } \mathbf{c}^{\mathrm{T}}\mathbf{x} \\ \text{subject to } \mathbf{A}\mathbf{x} \geq \mathbf{b} \end{array}} \quad (3.2)$$

As an illustrative example, consider a *diet problem* whose objective is to minimize the cost of a diet subject to a given nutritional requirement. Suppose that there are only two food items whose costs and nutritional values are given below.

|  | *apple juice* | *orange juice* | *minimum intake (unit)* |
|---|:---:|:---:|:---:|
| vitamin A (unit per glass) | 1 | 2 | 2 |
| vitamin B (unit per glass) | 2 | 1 | 2 |
| cost (unit per glass) | 3 | 1 | |

---

[1] In the majority of references on linear optimization [e.g. Bertsimas and Tsitsiklis, 1997; Papadimitriou and Steiglitz, 1998], the inequality constraints are written with "≥" instead of with "≤".

Let $x_1$ and $x_2$ be the amounts (in glass) of apple juice and orange juice respectively. The associated linear optimization problem is shown below.

$$\text{minimize } 3x_1 + x_2$$
$$\text{subject to } x_1 + 2x_2 \geq 2$$
$$2x_1 + x_2 \geq 2$$
$$x_1, x_2 \geq 0 \tag{3.3}$$

The feasible set is illustrated by the shaded region in figure 3.1.

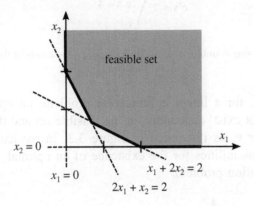

**Figure 3.1**  The feasible set of the example diet problem.

An optimal solution in this example can be found using the graphical approach as illustrated in figure 3.2. In particular, several contour lines are drawn. Note that $\mathbf{c} = (3, 1)$ is the direction of cost increase. From the contour lines, it is easy to see that the unique optimal solution is at the corner point $(0, 2)$ with the optimal cost equal to 2.

It should be noted that the cost vector $\mathbf{c}$ is equal to the gradient of the objective function, and is always perpendicular to any countour line in $\mathbb{R}^2$. To see this, consider two points $\mathbf{x}_1$ and $\mathbf{x}_2$ on the same countour line. From $\mathbf{c}^T\mathbf{x}_1 = \mathbf{c}^T\mathbf{x}_2$, we can write

$$\mathbf{c}^T(\mathbf{x}_1 - \mathbf{x}_2) = 0.$$

Since $\mathbf{x}_1 - \mathbf{x}_2$ is tangent to the contour line, it follows that $\mathbf{c}$ is perpendicular to the contour line.

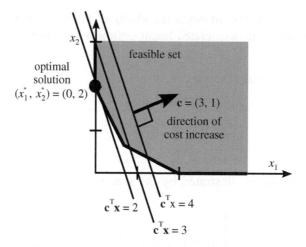

**Figure 3.2**    Using countour lines to find an optimal solution of the example diet problem.

In general, for a linear optimization problem, an optimal solution may or may not exist depending on the feasible set and the direction of the cost vector **c**, as illustrated in figure 3.3. In particular, figure 3.3 shows three possibilities for the existence of an optimal solution for a linear optimization problem.

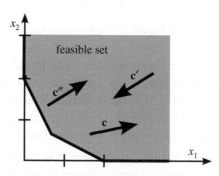

**Figure 3.3**    Effects of cost vector **c** on the existence of an optimal solution.

1.  There exists a unique optimal solution at some corner point of the feasible set. This case is represented by **c**.
2.  There is no optimal solution since the optimal cost is $-\infty$. This case is represented by **c**′.

3. There are infinitely many optimal solutions. This case is represented by $\mathbf{c}''$. Note that the contour lines for $\mathbf{c}''$ are parallel with one edge of the feasible set.

From the above illustrative example, one may guess that an optimal solution (if it exists) can be found at some corner point (if it exists) of the feasible set. This property is indeed valid for a linear optimization problem, and will be exploited to construct an efficient solution algorithm in the following sections.

## 3.2   Properties of Linear Optimization Problems

This section presents formally basic properties of linear optimization problems. In what follows, assume that $\mathbf{x}$, $\mathbf{c} \in \mathbb{R}^N$, $\mathbf{b} \in \mathbb{R}^M$, and $\mathbf{A} \in \mathbb{R}^{MN}$. Note that this assumption is consistent with the general form in (3.2).

### Polyhedrons and Extreme Points

For $\mathbf{a} \in R^N$ and $b \in \mathbb{R}$, a subset of $\mathbb{R}^N$ of the form $\{\mathbf{x} \in \mathbb{R}^N | \mathbf{a}^T\mathbf{x} = b\}$ is called a *hyperplane*. A subset of $\mathbb{R}^N$ of the form $\{\mathbf{x} \in \mathbb{R}^N | \mathbf{a}^T\mathbf{x} \geq b\}$ is called a *halfspace*.

A *polyhedron* is an intersection of a finite number of halfspaces. For $\mathbf{A} \in \mathbb{R}^{MN}$ and $\mathbf{b} \in \mathbb{R}^M$, note that a subset of $\mathbb{R}^N$ of the form $\{\mathbf{x} \in \mathbb{R}^N | \mathbf{A}\mathbf{x} \geq \mathbf{b}\}$ is a polyhedron. In particular, the feasible set of the linear optimization problem in (3.2) is a polyhedron.

It is straightforward to show that a polyhedron is a convex set. Since a linear cost function is convex, linear optimization is a special case of convex optimization. Therefore, all the properties that have been established for convex optimization are also valid for linear optimization.

A point $\mathbf{x}$ in a polyhedron is an *extreme point* if it is not a convex combination of any two distinct points in the same polyhedron. An extreme point of a polyhedron may not exist. Figure 3.4 shows some examples of extreme points. Note that an extreme point is previously referred to informally as a corner point.

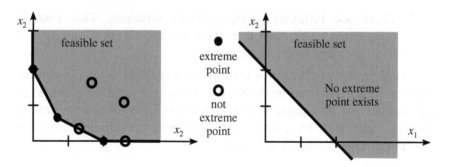

**Figure 3.4**   Examples of extreme points

## Standard Form of Linear Optimization

Instead of the general form in (3.2), consider a linear optimization problem in the *standard form* given below.

$$
\begin{aligned}
&\text{minimize } \mathbf{c}^T\mathbf{x} \\
&\text{subject to } \mathbf{A}\mathbf{x} = \mathbf{b} \\
&\qquad\qquad \mathbf{x} \geq \mathbf{0}
\end{aligned}
\tag{3.4}
$$

Any general-form problem can be transformed into a standard-form problem. The transformation consists of the following two steps.

1.  Elimination of unrestricted variables: An *unrestricted* variable can be positive, zero, or negative. To transform a general-form problem into a standard-form problem, an unrestricted variable $x_i$ is replaced with two nonnegative variables $x_i^+, x_i^- \geq 0$ by setting $x_i = x_i^+ - x_i^-$. This transformation is based on a simple observation that any real number can be written as the difference between two nonnegative numbers, e.g. $2 = 2 - 0$ and $-2 = 0 - 2$.

2.  Elimination of inequality constraints. Constraints $\mathbf{A}\mathbf{x} \geq \mathbf{b}$ can be expressed as

$$
\mathbf{A}\mathbf{x} - \mathbf{z} = \mathbf{b} \text{ and } \mathbf{z} \geq \mathbf{0},
$$

where $\mathbf{z}$ contains additional variables referred to as *surplus variables*. Note that $\mathbf{z} \in \mathbb{R}^M.$[2]

---

[2]Similarly, constraints of the form $\mathbf{A}\mathbf{x} \leq \mathbf{b}$ can be expressed as $\mathbf{A}\mathbf{x} + \mathbf{z} = \mathbf{b}$ and $\mathbf{z} \geq \mathbf{0}$, where $\mathbf{z}$ contains additional variables referred to as *slack variables*. However, slack variables are not needed in the transformation from the problem in (3.2).

Note that, for a problem in the standard form, the feasible set is a polyhedron. To see this, observe that each hyperplane $\mathbf{a}^T\mathbf{x} = b$ can be viewed as an intersection of two halfspaces $\mathbf{a}^T\mathbf{x} \geq b$ and $\mathbf{a}^T\mathbf{x} \leq b$.

**Example 3.1:** Consider the diet problem in (3.3). The corresponding standard-form problem is shown below.

$$\text{minimize } 3x_1 + x_2 \qquad\qquad \text{minimize } 3x_1 + x_2$$
$$\text{subject to } x_1 + 2x_2 \geq 2 \quad \Rightarrow \quad \text{subject to } x_1 + 2x_2 - z_1 = 2$$
$$2x_1 + x_2 \geq 2 \qquad\qquad 2x_1 + x_2 - z_2 = 2$$
$$x_1, x_2 \geq 0 \qquad\qquad x_1, x_2, z_1, z_2 \geq 0$$

Below is another example of transforming a problem into a standard-form problem.

$$\text{minimize } 3x_1 + x_2 \qquad\qquad \text{minimize } 3x_1^+ - 3x_1^- + x_2^+ - x_2^-$$
$$\text{subject to } x_1 + 2x_2 = 2 \Rightarrow \text{subject to } x_1^+ - x_1^- + 2x_2^+ - 2x_2^- = 2$$
$$2x_1 + x_2 \geq 2 \qquad\qquad 2x_1^+ - 2x_1^- + x_2^+ - x_2^- - z_1 = 2$$
$$x_1^+, x_1^-, x_2^+, x_2^-, z_1 \geq 0$$

Notice that both the number of variables and the number of constraints may change as a result of the problem transformation. □

Without loss of generality, the constraint matrix $\mathbf{A}$ in (3.4) can be assumed to have full rank, i.e. $\text{rank}(\mathbf{A}) = M$. This assumption is justified by the following theorem.

**Theorem 3.1 (Justification of full-rank assumption for standard-form problems)** Suppose that the feasible set $\mathcal{F} = \{\mathbf{x} \in \mathbb{R}^N | \mathbf{A}\mathbf{x} = \mathbf{b}, \mathbf{x} \geq 0\}$ is nonempty. In addition, suppose that $\text{rank}(\mathbf{A}) = K < M$, and that rows $\mathbf{a}_{i_1}^T, ..., \mathbf{a}_{i_K}^T$ of $\mathbf{A}$ are linearly independent. Then, $\mathcal{F}$ is the same as the polyhedron $\mathcal{G} = \{\mathbf{x} \in \mathbb{R}^N | \mathbf{a}_{i_1}^T\mathbf{x} = b_{i_1}, ..., \mathbf{a}_{i_K}^T\mathbf{x} = b_{i_K}, \mathbf{x} \geq 0\}$.

**Proof** Since any point in $\mathcal{F}$ satisfies all the constraints in the definition of $\mathcal{G}$, it is clear that $\mathcal{F} \subset \mathcal{G}$. It remains to show that $\mathcal{G} \subset \mathcal{F}$.

It can be assumed without loss of generality that $(i_1, ..., i_K) = (1, ..., K)$. This condition can be achieved by rearranging the order of constraints. Since $\text{rank}(\mathbf{A}) = K$, any row $\mathbf{a}_i^T$ can be written as a linear combination $\mathbf{a}_i^T = \sum_{k=1}^K \alpha_{ik} \mathbf{a}_k^T$ for some scalars $\alpha_{i1}, ..., \alpha_{iK}$.

Consider now an arbitrary point $\mathbf{x} \in \mathcal{F}$. This point can be used to write

$$b_i = \mathbf{a}_i^T \mathbf{x} = \sum_{k=1}^{K} \alpha_{ik} \, \mathbf{a}_k^T \, \mathbf{x} = \sum_{k=1}^{K} \alpha_{ik} \, b_k, \; i \in \{1, \ldots, M\}.$$

Consider any point $\mathbf{y} \in \mathcal{G}$. From the above expression, i.e. $b_i = \sum_{k=1}^{K} \alpha_{ik} \, b_k$,

$$\mathbf{a}_i^T \mathbf{y} = \sum_{k=1}^{K} \alpha_{ik} \, \mathbf{a}_k^T \, \mathbf{y} = \sum_{k=1}^{K} \alpha_{ik} \, b_k = b_i, \; i \in \{1, \ldots, M\},$$

which implies that any point $\mathbf{y} \in \mathcal{G}$ belongs to $\mathcal{F}$. Thus, $\mathcal{G} \subset \mathcal{F}$. □

## Basic Feasible Solutions

Let $\mathbf{A}_i$ denote column $i$ of the constraint matrix $\mathbf{A}$. Define a *basis* of $\mathbf{A}$ as a set of $M$ linearly independent columns of $\mathbf{A}$. A basis can be expressed as an $M \times M$ nonsingular matrix $\mathbf{B} = [\mathbf{A}_{B(1)} \ldots \mathbf{A}_{B(M)}]$. For convenience, let $\mathcal{B} = \{B(1), \ldots, B(M)\}$ and $\mathbb{Z}_M = \{1, \ldots, M\}$.

Define the *basic solution* corresponding to $\mathbf{B}$ to be a solution $\mathbf{x} \in \mathbb{R}^N$ such that

1. $x_i = 0$ for all $i \notin \mathcal{B}$

2. $x_{B(i)} = i$th component of $\mathbf{B}^{-1}\mathbf{b}$ for $i \in \mathbb{Z}_M$.

Given a basis $\mathbf{B} = [\mathbf{A}_{B(1)} \ldots \mathbf{A}_{B(M)}]$, the components $x_{B(1)}, \ldots, x_{B(M)}$ of $\mathbf{x}$ are called *basic variables*. A basic solution need not be feasible. If a basic solution is feasible, it is called a *basic feasible solution* (*BFS*).

**Example 3.2:** Consider the following linear optimization problem.

$$\text{minimize } x_1 + 2x_2 + 3x_3$$

$$\text{subject to } \begin{bmatrix} 1 & 1 & 1 & 1 & 0 & 0 \\ 1 & 0 & 0 & 0 & 1 & 0 \\ 0 & 1 & 2 & 0 & 0 & 1 \end{bmatrix} \begin{bmatrix} x_1 \\ x_2 \\ x_3 \end{bmatrix} = \begin{bmatrix} 2 \\ 3 \\ 4 \end{bmatrix}$$

$$x_1, x_2, x_3 \geq 0$$

One basis is $\mathbf{B} = [\mathbf{A}_4 \, \mathbf{A}_5 \, \mathbf{A}_6]$, which is the identity matrix. The corresponding basic solution is $\mathbf{x} = (0, 0, 0, 2, 3, 4)$. Since $\mathbf{x}$ is feasible, it is a BFS. Another basis is $\mathbf{B}' = [\mathbf{A}_1 \, \mathbf{A}_3 \, \mathbf{A}_4]$. The corresponding basic solution is $\mathbf{x}' = (3, 0, 2, -3, 0, 0)$, which is not a BFS.

The above optimization problem also demonstrates that two different bases can have the same BFS. Consider the bases $\mathbf{B}'' = [\mathbf{A}_1 \, \mathbf{A}_3 \, \mathbf{A}_5]$ and $\mathbf{B}''' = [\mathbf{A}_3 \, \mathbf{A}_4 \, \mathbf{A}_5]$. Both bases have the same BFS equal to $(0, 0, 2, 0, 3, 0)$. □

For a BFS to correspond to more than one basis, it must contain more than $N–M$ zero components. A BFS that contains more than $N–M$ zero components is called *degenerate*. The following theorem states that, for a nonempty feasible set $\mathcal{F}$, a BFS is equivalent to an extreme point of $\mathcal{F}$. The proof is adapted from [Papadimitriou and Steiglitz, 1998].

**Theorem 3.2 (Equivalence between an extreme point and a BFS)**   If the feasible set $\mathcal{F}$ of the standard-form problem in (3.4) is nonempty, then $\mathbf{x}^* \in \mathcal{F}$ is a BFS if and only if it is an extreme point of $\mathcal{F}$.

The proof of theorem 3.2 relies on the following lemma.

**Lemma 3.1 [Papadimitriou and Steiglitz, 1998]**   If $\mathbf{x}^*$ is a BFS in $\mathcal{F}$, where $\mathcal{F}$ is the feasible set of the standard-form problem in (3.4), then there is a cost vector $\mathbf{c}$ such that $\mathbf{x}^*$ is the unique optimal solution to the problem of minimizing $\mathbf{c}^\mathrm{T}\mathbf{x}$ subject to $\mathbf{x} \in \mathcal{F}$.

[†]**Proof**   Let $\mathcal{B} = \{B(1), ..., B(M)\}$ be the set of indices for the basic variables in $\mathbf{x}^*$. Define $\mathbf{c}$ as follows.

$$c_i = \begin{cases} 0, & i \in \mathcal{B} \\ 1, & \text{otherwise} \end{cases}$$

Since $\mathbf{c} \geq \mathbf{0}$, the optimal cost is at least zero. Since $x_i^* = 0$ for $i \notin \mathcal{B}$, $\mathbf{c}^\mathrm{T}\mathbf{x}^* = 0$. It follows that $\mathbf{x}^*$ is optimal. In addition, any optimal solution must satisfy $x_i = 0$ for $i \notin \mathcal{B}$, and thus $\mathbf{B}\mathbf{x}_B = \mathbf{b}$, where $\mathbf{B} = [A_{B(1)}, ..., A_{B(M)}]$ and $\mathbf{x}_B = (x_{B(1)}, ..., x_{B(M)})$. Since $\mathbf{B}$ is invertible, $\mathbf{x}^*$ obtained from $\mathbf{x}_B = \mathbf{B}^{-1}\mathbf{b}$ and $x_i = 0$ for $i \notin \mathcal{B}$ is the unique optimal solution.

[†]**Proof of theorem 3.2**   Suppose that $\mathbf{x}^*$ is an extreme point of $\mathcal{F}$. To show that $\mathbf{x}^*$ is a BFS, consider the set of columns of $\mathbf{A}$ defined by $\mathcal{A}' = \{\mathbf{A}_i | x_i^* > 0\}$. The columns in $\mathcal{A}'$ are linearly independent, as shown below by contradiction.

Let $\mathcal{B}'$ be the set of column indices in $\mathcal{A}'$. Suppose that the columns in $\mathcal{A}'$ are linearly dependent. Then, there are numbers $d_i$'s, not all zero, such that $\sum_{i \in \mathcal{B}'} d_i \mathbf{A}_i = \mathbf{0}$. Since $\mathbf{x}^*$ is feasible, $\mathbf{A}\mathbf{x}^* = \mathbf{b}$ or equivalently $\sum_{i \in \mathcal{B}'} x_i^* \mathbf{A}_i = \mathbf{b}$. Multiplying $\sum_{i \in \mathcal{B}'} d_i \mathbf{A}_i = \mathbf{0}$ by $\pm \theta$ for $\theta > 0$ and adding the result to $\sum_{i \in \mathcal{B}'} x_i^* \mathbf{A}_i = \mathbf{b}$ yields

$$\sum_{i \in \mathcal{B}'} (x_i^* \pm \theta d_i) \mathbf{A}_i = \mathbf{b}.$$

Since $x_i > 0$ for $i \in \mathcal{B}'$, there is a small $\theta > 0$ such that $x_i^* \pm \theta d_i \geq 0$ for $i \in \mathcal{B}'$. The two points $\mathbf{x}'$ and $\mathbf{x}''$ defined by

$$x_i' = \begin{cases} x_i^* - \theta d_i & i \in \mathcal{B}' \\ 0 & \text{otherwise} \end{cases} \quad \text{and} \quad x_i'' = \begin{cases} x_i^* + \theta d_i & i \in \mathcal{B}' \\ 0 & \text{otherwise} \end{cases}$$

satisfy $\mathbf{A}\mathbf{x}' = \mathbf{A}\mathbf{x}'' = \mathbf{b}$ and $\mathbf{x}', \mathbf{x}'' \geq \mathbf{0}$. Thus, $\mathbf{x}', \mathbf{x}'' \in \mathcal{F}$. Since $\mathbf{x}^* = \frac{1}{2}\mathbf{x}' + \frac{1}{2}\mathbf{x}''$, $\mathbf{x}^*$ is not an extreme point, which is a contradiction.

Since the columns in $\mathcal{A}'$ are linearly independent, $|\mathcal{B}'| \leq M$. Hence, $\mathcal{A}'$ can be augmented if necessary to form a basis with $M$ linearly independent columns, implying that $\mathbf{x}^*$ is a BFS.

For the converse, suppose that $\mathbf{x}^*$ is a BFS. From lemma 3.1, there is a cost vector $\mathbf{c}$ such that $\mathbf{x}^*$ is the unique point in $\mathcal{F}$ that minimizes $\mathbf{c}^T\mathbf{x}$. In other words, $\mathbf{c}^T\mathbf{x} > \mathbf{c}^T\mathbf{x}^*$ for all $\mathbf{x} \in \mathcal{F}$ with $\mathbf{x} \neq \mathbf{x}^*$. To show that $\mathbf{x}^*$ is an extreme point of $\mathcal{F}$, contradiction is used below.

Suppose that $\mathbf{x}^*$ is not an extreme point. Then, $\mathbf{x}^* = \alpha \mathbf{x}' + (1 - \alpha)\mathbf{x}''$ for some $\mathbf{x}', \mathbf{x}'' \in \mathcal{F}$ and $\alpha \in (0, 1)$. Since $\mathbf{x}^*$ is the unique minimum point of $\mathbf{c}^T\mathbf{x}$, it follows that

$$\mathbf{c}^T\mathbf{x}^* = \alpha^*\mathbf{c}^T\mathbf{x}' + (1 - \alpha)\mathbf{c}^T\mathbf{x}'' > \alpha\mathbf{c}^T\mathbf{x}^* + (1 - \alpha)\mathbf{c}^T\mathbf{x}^* = \mathbf{c}^T\mathbf{x}^*,$$

which is a contradiction. □

The next theorem states that a linear optimization problem in the standard form has at least one BFS if its feasible set is not empty.

**Theorem 3.3 (Existence of a BFS)** Consider the linear optimization problem in (3.4). If the feasible set $\mathcal{F}$ is not empty, then $\mathcal{F}$ has at least one BFS.

**Proof** Consider an arbitrary point $\mathbf{x} \in \mathcal{F}$. Let $M'$ be the number of positive components of $\mathbf{x}$, and $\mathcal{B}' = \{B'(1), ..., B'(M')\}$ be the set of their indices. If the columns $\mathbf{A}_{B'(1)}, ..., \mathbf{A}_{B'(M')}$ are linearly independent, they can be augmented if necessary to obtain a basis, implying that $\mathbf{x}$ is a BFS. It remains to consider $\mathbf{A}_{B'(1)}, ..., \mathbf{A}_{B'(M')}$ that are linearly dependent.

Since $\mathbf{A}_{B'(1)}, ..., \mathbf{A}_{B'(M')}$ are linearly dependent, there are numbers $d_i$'s, not all zero, such that $\sum_{i \in \mathcal{B}'} d_i \mathbf{A}_i = \mathbf{0}$. Since $\mathbf{x}^*$ is feasible, $\mathbf{A}\mathbf{x}^* = \mathbf{b}$ or equivalently $\sum_{i \in \mathcal{B}'} x_i^* \mathbf{A}_i = \mathbf{b}$. Multiplying $\sum_{i \in \mathcal{B}'} d_i \mathbf{A}_i = \mathbf{0}$ by $\theta > 0$ and adding the result to $\sum_{i \in \mathcal{B}'} x_i^* \mathbf{A}_i = \mathbf{b}$ yields

$$\sum_{i \in \mathcal{B}'} (x_i^* + \theta d_i) \mathbf{A}_i = \mathbf{b}.$$

Without loss of generality, assume that some $d_i$'s are negative. (If this is not the case, simply replace the original $d_i$'s with their negatives.) Since $x_i > 0$ for $i \in \mathcal{B}'$, setting $\theta$ equal to $\theta^* = \min_{i \in \mathcal{B}', d_i < 0}(-x_i/d_i)$ yields $\mathbf{x}' \in \mathcal{F}$ with the number of positive components smaller than $M'$.

In summary, given that $\mathbf{A}_{B'(1)}, ..., \mathbf{A}_{B'(M')}$ are linearly dependent, there is another feasible solution with a smaller number of positive components. Repeating the process will eventually yield a feasible solution whose columns $\mathbf{A}_{B'(1)}, ..., \mathbf{A}_{B'(M')}$ are linearly independent, yielding a BFS.

## Fundamental Theorem of Linear Optimization

Consider a linear optimization problem in the standard form that is feasible. One specific property of linear optimization is that, unless the optimal cost is $-\infty$, there exists at least one optimal solution with a finite optimal cost. This property is not valid for nonlinear optimization. For example, consider minimizing $1/x$ subject to $x \geq 0$. Although there is no solution with cost $-\infty$, there is no optimal solution. The property is stated formally in the following theorem.

**Theorem 3.4 (Fundamental theorem of linear optimization)** For the linear optimization problem in (3.4) that is feasible, either the optimal cost is $-\infty$ or there exists a BFS that is an optimal solution.

**Proof** It suffices to consider a feasible problem whose optimal cost is not $-\infty$. It is shown below that, for any $\mathbf{x} \in \mathcal{F}$, there is a BFS $\mathbf{w}$ with $\mathbf{c}^T\mathbf{w} \leq \mathbf{c}^T\mathbf{x}$. This statement will imply that, among all the BFSs, a BFS with the minimum cost, denoted by $\mathbf{w}^*$, is an optimal solution to the problem, i.e. $\mathbf{c}^T\mathbf{w} \leq \mathbf{c}^T\mathbf{x}$ for all $\mathbf{x} \in \mathcal{F}$.

Consider an arbitrary $\mathbf{x} \in \mathcal{F}$. Let $M'$ be the number of positive components of $\mathbf{x}$, and let $\mathcal{B}' = \{B'(1), \dots, B'(M')\}$ be the set of their indices. If the columns $\mathbf{A}_{B'(1)}, \dots, \mathbf{A}_{B'(M')}$ are linearly independent, they can be augmented if necesssary to obtain a basis, implying that $\mathbf{x}$ is a BFS and thus $\mathbf{w}$ can be set equal to $\mathbf{x}$. It remains to consider $\mathbf{A}_{B'(1)}, \dots, \mathbf{A}_{B'(M')}$ that are linearly dependent.

Since $\mathbf{A}_{B'(1)}, \dots, \mathbf{A}_{B'(M')}$ are linearly dependent, there are numbers $d_i$'s, not all zero, such that $\sum_{i \in B'} d_i \mathbf{A}_i = \mathbf{0}$. Since $\mathbf{x}$ is feasible, $\mathbf{A}\mathbf{x} = \mathbf{b}$ or equivalently $\sum_{i \in B'} x_i \mathbf{A}_i = \mathbf{b}$. Multiplying $\sum_{i \in B'} d_i \mathbf{A}_i = \mathbf{0}$ by a real number $\theta$ and adding the result to $\sum_{i \in B'} x_i \mathbf{A}_i = \mathbf{b}$ yields

$$\sum_{i \in B'} (x_i + \theta d_i)\mathbf{A}_i = \mathbf{b}.$$

Without loss of generality, assume that $\sum_{i \in B'} c_i d_i \leq 0$. (If not, $d_i$ can be replaced by $-d_i$ for all $i \in \mathcal{B}'$.) Consider two separate cases.

1. $\sum_{i \in B'} c_i d_i < 0$: In this case, the higher the value of $\theta$, the smaller the cost of a new solution obtained from changing $x_i$ to $x_i + \theta d_i$ for $i \in \mathcal{B}'$. However, since the optimal cost is not $-\infty$, there exists a maximum value of $\theta$, denoted by $\theta^*$, such that $x_i + \theta^* d_i = 0$ for some $i \in \mathcal{B}'$; increasing $\theta$ beyond $\theta^*$ yields an infeasible solution due to the non-negativity constraint, i.e. $\mathbf{x} \geq \mathbf{0}$.

2. $\sum_{i \in B'} c_i d_i = 0$: In this case, there is no cost change from changing $x_i$ to $x_i + \theta d_i$ for $i \in \mathcal{B}'$. Due to the non-negativity constraint, as

$|\theta|$ increases, $x_i + \theta d_i = 0$ for some $i \in \mathcal{B}'$. Let $\theta^*$ denote the value of $\theta$ with the minimum $|\theta|$ such that $x_i + \theta d_i = 0$ for some $i \in \mathcal{B}'$. Note that $\theta^*$ can be either positive or negative.

In either case, another feasible solution $\mathbf{x}'$ can be obtained as shown below. Compared to $\mathbf{x}$, the cost of $\mathbf{x}'$ can only decrease while the number of positive components is strictly smaller.

$$x_i' = \begin{cases} x_i + \theta^* d_i & i \in \mathcal{B}' \\ 0, & \text{otherwise} \end{cases}$$

Repeating the process will eventually yield a feasible solution whose columns $\mathbf{A}_{B'(1)}$, ..., $\mathbf{A}_{B'(M')}$ are linearly independent, yielding a BFS $\mathbf{w}$ with no higher cost than $\mathbf{x}$. $\qquad\Box$

Theorem 3.4 implies that, to find an optimal solution (when it exists), it is sufficient to check all the BFSs, or equivalently all the extreme points of the feasible set. This property gives rise to the well known simplex algorithm, which systematically checks the set of extreme points of the feasible set for an optimal solution. The simplex algorithm is discussed in detail in the next section.

## 3.3 Simplex Algorithm

Consider the standard-form problem in (3.4). The *simplex algorithm* provides a systematic procedure in moving from one BFS to another such that, at each iteration, the next BFS is no worse than the current BFS in terms of the objective function value.

Suppose that the current BFS $\mathbf{x}$ has basis $\mathbf{B} = [\mathbf{A}_{B(1)}, ..., \mathbf{A}_{B(M)}]$. Recall that $\mathcal{B} = \{B(1), ..., B(M)\}$ are the set of indices for basic variables, i.e. $x_{B(1)}, ..., x_{B(M)}$. In what follows, the columns $\mathbf{A}_{B(1)}, ..., \mathbf{A}_{B(M)}$ are referred to as *basic columns*.

For a feasible solution $\mathbf{x} \in \mathcal{F}$, a vector $\mathbf{d}$ is a *feasible direction* at $\mathbf{x}$ if there is a $\theta > 0$ such that $\mathbf{x} + \theta\mathbf{d} \in \mathcal{F}$. Consider moving away from $\mathbf{x}$ along a feasible direction $\mathbf{d}^j$ to $\mathbf{x} + \theta\mathbf{d}^j$ such that one nonbasic variable $x_j$ increases from 0 to $\theta > 0$. This implies that $d_j^j = 1$ while $d_i^j = 0$ for all $i \notin \mathcal{B} \cup \{j\}$. Since $\mathbf{x} + \theta\mathbf{d}^j$ is feasible, it follows that $\mathbf{A}(\mathbf{x} + \theta\mathbf{d}^j) = \mathbf{b}$, yielding $\mathbf{A}\mathbf{d}^j = \mathbf{0}$ since $\mathbf{A}\mathbf{x} = \mathbf{b}$.

For convenience, define $\mathbf{x}_B = (x_{B(1)}, \ldots, x_{B(M)})$, $\mathbf{c}_B = (c_{B(1)}, \ldots, c_{B(M)})$, and $\mathbf{d}_B^j = (d_{B(1)}^j, \ldots, d_{B(M)}^j)$, to refer to the components of $\mathbf{x}$, $\mathbf{c}$, and $\mathbf{d}$ corresponding to the basic variables. From the condition $\mathbf{A}\mathbf{d}^j = \mathbf{0}$,

$$\mathbf{0} = \sum_{i=1}^{N} \mathbf{A}_i d_i^j = \sum_{i \in B} \mathbf{A}_i d_i^j + \mathbf{A}_j = \mathbf{B}\mathbf{d}_B^j + \mathbf{A}_j,$$

yielding $\mathbf{d}_B^j = -\mathbf{B}^{-1}\mathbf{A}_j$. The vector $\mathbf{d}^j$ whose value is defined by $\mathbf{d}_B^j = -\mathbf{B}^{-1}\mathbf{A}_j$, $d_j^j = 1$, and $d_i^j = 0$ for $i \notin B \cup \{j\}$, is called the *basic direction* for nonbasic variable $x_j$.

While moving from $\mathbf{x}$ along $\mathbf{d}^j$, the rate of cost change along this direction is

$$\mathbf{c}^T\mathbf{d}^j = \sum_{i=1}^{N} c_i d_i^j = \sum_{i \in B} c_i d_i^j + c_j d_j^j = \mathbf{c}_B^T \mathbf{d}_B^j + c_j = c_j - \mathbf{c}_B^T \mathbf{B}^{-1}\mathbf{A}_j.$$

Let $\bar{c}_j = c_j - \mathbf{c}_B^T\mathbf{B}^{-1}\mathbf{A}_j$. The quantity $\bar{c}_j$ is called the *reduced cost* of nonbasic variable $x_j$. Roughly speaking, the reduced cost $\bar{c}_j$ is the cost change per unit of movement along the basic direction of $x_j$.

The definition of the basic direction and the reduced cost can be extended for basic variables. In particular, for basic variable $x_j$, the basic direction $\mathbf{d}^j$ is the unit vector with the $j$th component equal to 1. Note that this definition is consistent with the previously stated conditions: $\mathbf{d}_B^j = -\mathbf{B}^{-1}\mathbf{A}_j$, $d_j^j = 1$, and $d_i^j = 0$ for all $i \notin B \cup \{j\}$.

It follows that the reduced cost expression $\bar{c}_j = c_j - \mathbf{c}_B^T\mathbf{B}^{-1}\mathbf{A}_j$ is also valid for basic variable $x_j$. However, for any basic variable $x_j$, the reduced cost $\bar{c}_j$ is always zero. To see this, note that $\mathbf{B}^{-1}\mathbf{A}_{B(i)} = \mathbf{e}_i$, where $\mathbf{e}_i$ is the unit vector with the $i$th component equal to 1. Therefore,

$$\bar{c}_{B(i)} = c_{B(i)} - \mathbf{c}_B^T\mathbf{B}^{-1}\mathbf{A}_{B(i)} = c_{B(i)} - \mathbf{c}_B^T\mathbf{e}_i = c_{B(i)} - c_{B(i)} = 0.$$

Le $\bar{\mathbf{c}} = (\bar{c}_1, \ldots, \bar{c}_N)$. If a BFS is found with all reduced costs being nonnegative, i.e. $\bar{\mathbf{c}} \geq \mathbf{0}$, it means that the cost cannot be decreased further by moving away from this BFS. As a result, the reduced costs are important quantities that indicate the optimality of a BFS, as stated formally below.

**Theorem 3.5 (Optimality condition of a BFS)** Consider a BFS $\mathbf{x}^*$ associated with basis $\mathbf{B}$.

1. If $\bar{\mathbf{c}} \geq \mathbf{0}$, then $\mathbf{x}^*$ is an optimal solution.
2. If $\mathbf{x}^*$ is optimal and nondegenerate, then $\bar{\mathbf{c}} \geq \mathbf{0}$.

**Proof**

1. Suppose that $\bar{\mathbf{c}} \geq \mathbf{0}$, which is equivalent to $\mathbf{c}^T \geq \mathbf{c}_B^T \mathbf{B}^{-1} \mathbf{A}$. Consider an arbitrary $\mathbf{y} \in \mathcal{F}$. From $\bar{\mathbf{c}} \geq \mathbf{0}$ and $\mathbf{A}\mathbf{y} = \mathbf{b}$,

$$\mathbf{c}^T \mathbf{y} \geq \mathbf{c}_B^T \mathbf{B}^{-1} \mathbf{A}\mathbf{y} = \mathbf{c}_B^T \mathbf{B}^{-1} \mathbf{b} = \mathbf{c}_B^T \mathbf{x}_B = \mathbf{c}^T \mathbf{x}^*,$$

which implies that $\mathbf{x}^*$ is optimal.

2. Assume that $\mathbf{x}^*$ is optimal and nondegenerate. It will be shown by contradiction that $\bar{\mathbf{c}} \geq \mathbf{0}$.

Suppose that $\bar{c}_j < 0$ for some $j$. Since the reduced cost is zero for a basic variable, $x_j^*$ is nonbasic. Since $\mathbf{x}^*$ is nondegenerate, all basic variables are positive. Using the basic direction $\mathbf{d}_j$ for $x_j^*$, for a sufficiently small $\theta > 0$, $\mathbf{x}^* + \theta \mathbf{d}^j$ is feasible, i.e. nonnegative, and has a lower cost than $\mathbf{x}^*$, contradicting the assumption that $\mathbf{x}^*$ is optimal. $\square$

## Moving from One BFS to Another

In cases where the reduced cost is negative, it is possible to move from the current BFS to another with a lower cost. The simplex algorithm specifies how to do so systematically. Consider a BFS $\mathbf{x}$ with some $\bar{c}_j < 0$. Let $\mathcal{B}$ be the set of indices for basic variables. Consider moving from $\mathbf{x}$ along the basic direction $\mathbf{d}^j$ of nonbasic variable $x_j$. In particular, the new solution has the form $\mathbf{x} + \theta \mathbf{d}^j$ with $\theta > 0$. An immediate question is how to set the value of $\theta$. There are two possibilities.

If $\mathbf{d}^j \geq \mathbf{0}$, $\theta$ can be increased indefinitely without violating the non-negativity constraint. In this case, it can be concluded that the optimal cost is $-\infty$.

If $d_i^j < 0$ for some $i$, then $\theta$ can be increased only up to the following amount.

$$\theta^* = \min_{i \in \mathcal{B}, d_i^j < 0} \left( -\frac{x_i}{d_i^j} \right) \tag{3.5}$$

For convenience, define $i^* = \arg\min_{i \in B, d_i^j < 0} (-x_i/d_i^j)$. It follows that moving from $\mathbf{x}$ to $\mathbf{x} + \theta^* \mathbf{d}^j$ results in $x_{i*} = 0$. The next theorem states that $\mathbf{x} + \theta^* \mathbf{d}^j$ is in fact another BFS. Therefore, the described process can move from one BFS to another BFS.

**Theorem 3.6**   Given that $\mathbf{x}$ is a BFS, the solution $\mathbf{x} + \theta^* \mathbf{d}^j$ where $\mathbf{d}^j$ is the basic direction for nonbasic variable $x_j$ and $\theta^* = \min_{i \in B, d_i^j < 0}(-x_i/d_i^j)$, is another BFS. The new basic columns are $\mathbf{A}_j$ and $\mathbf{A}_i$, $i \in B - \{i^*\}$, where $i^* = \arg\min_{i \in B, d_i^j < 0} (-x_i/d_i^j)$.

**Proof**   Let $B' = B \cup \{j\} - \{i^*\}$ denote the new set of indices for basic variables. The first part of the proof is to establish that columns $\mathbf{A}_i$, $i \in B'$, are linearly independent. Consider a set of numbers $\alpha_i$'s such that

$$\sum_{i \in B'} \alpha_i \mathbf{A}_i = \alpha_j \mathbf{A}_j + \sum_{i \in B' - \{i^*\}} \alpha_i \mathbf{A}_i = \mathbf{0}. \qquad (3.6)$$

Recall that $\mathbf{d}^j$ satisfies $\mathbf{A}\mathbf{d}^j = \mathbf{0}$, $d_j^j = 1$, and $d_i^j = 0$ for $i \notin B \cup \{j\}$. Accordingly, $\mathbf{A}_j = -\sum_{i \in B} d_i^j \mathbf{A}^i$. Substituting this expression of $\mathbf{A}_j$ into (3.6) yields

$$-\alpha_j \sum_{i \in B} d_i^j \mathbf{A}_i + \sum_{i \in B' - \{i^*\}} \alpha_i \mathbf{A}_i = -\alpha_j d_{i*}^j \mathbf{A}_{i*} + \sum_{i \in B' - \{i^*\}} (\alpha_i - \alpha_j d_i^j) \mathbf{A}_i = \mathbf{0}.$$

Since $\mathbf{A}_i$, $i \in B$, are linearly independent, the coefficients in the above equality must all be zero. Thus, $\alpha_j d_{i*}^j = 0$ and $\alpha_i - \alpha_j d_i^j = 0$ for all $i \in B - \{i^*\}$. Since $d_{i*}^j < 0$ by the definition of $i^*$, $\alpha_j = 0$. Since $\alpha_i - \alpha_j d_i^j = 0$, it follows that $\alpha_i = 0$ for all $i \in B - \{i^*\}$. Thus, the coefficients $\alpha_i$'s in (3.6) are all zero, implying that $\mathbf{A}_i$, $i \in B'$, are linearly independent.

After moving from $\mathbf{x}$ to $\mathbf{x} + \theta^* \mathbf{d}^j$, $x_{i*} = 0$. In addition, the values of $x_i$, $i \notin B \cup \{j\}$, do not change. It follows that $x_i = 0$ for all $i \notin B'$.

In conclusion, since the basic columns $\mathbf{A}_i$, $i \in B'$, are linearly independent and the nonbasic variables $x_i$, $i \notin B'$, are all zero, the new solution $\mathbf{x} + \theta^* \mathbf{d}^j$ is a BFS.    □

The process of moving from one BFS to another is called *pivoting*. In the above discussion, column $\mathbf{A}_{i*}$ is said to *leave the basis* while column $\mathbf{A}_j$ *enters the basis*. If the value of $i^*$ is not unique, then the new BFS can correspond to more than one basis and is thus degenerate. In addition, it is possible to have $\theta^* = 0$ when pivoting from a degenerate BFS, as will be seen in examples. In this case, pivoting does not move the solution point but can change the basis.[3] Below is an iteration of the simplex algorithm.

## Iteration of Simplex Algorithm

1. Start with some BFS $\mathbf{x}$ with basis $\mathbf{B}$.
2. Compute the reduced cost $\bar{c}_j - \mathbf{c}_B^T \mathbf{B}^{-1} \mathbf{A}_j$ for all nonbasic variables. If all reduced costs are nonnegative, then the algorithm terminates by concluding that the current BFS is optimal. Otherwise, choose a nonbasic variable $x_j$ with $\bar{c}_j < 0$.
3. Compute the basic direction $\mathbf{d}^j$ for $x_j$ by setting $\mathbf{d}_B^j = -\mathbf{B}^{-1}\mathbf{A}_j$, $d_j^j = 1$, and $d_i^j = 0$ for $i \neq \mathcal{B} \cup \{j\}$.
4. If $\mathbf{d}^j \geq \mathbf{0}$, then the algorithm terminates by concluding that the optimal cost is $-\infty$. Otherwise, move from BFS $\mathbf{x}$ to BFS $\mathbf{x} + \theta^* \mathbf{d}^j$ with $\theta^* = \min_{i \in \mathcal{B}, \, d_i^j < 0} (-x_i/d_i^j)$.
5. Let $i^* = \arg\min_{i \in \mathcal{B}, \, d_i^j < 0} (-x_i/d_i^j)$. Make column $\mathbf{A}_{i^*}$ leave the basis and column $\mathbf{A}_j$ enter the basis. Go back to step 2 with the new BFS.

## Full Tableau Implementation

The *full tableau implementation* of the simplex algorithm relies on keeping a table of relevant information. In particular, the table is of the form

| $-\mathbf{c}_B^T \mathbf{B}^{-1}\mathbf{b}$ | $\mathbf{c}^T - \mathbf{c}_B^T \mathbf{B}^{-1}\mathbf{A}$ |
|---|---|
| $\mathbf{B}^{-1}\mathbf{b}$ | $\mathbf{B}^{-1}\mathbf{A}$ |

---

[3]However, two BFSs with the same solution point but with different bases are considered two different solutions.

or equivalently

| $-\mathbf{c}_B^T \mathbf{x}_B$ | $\overline{c}_1$ | ... | $\overline{c}_N$ |
|---|---|---|---|
| $x_{B(1)}$ $\vdots$ $x_{B(M)}$ | $\mathbf{B}^{-1}\mathbf{A}_1$ | ... | $\mathbf{B}^{-1}\mathbf{A}_N$ |

In each iteration, the left column or the *0th column* contains the basic variables $x_{B(1)}, ..., x_{B(M)}$. Note that column $\mathbf{B}^{-1}\mathbf{A}_{B(i)}$ is the unit vector $\mathbf{e}_i$ with its $i$th component equal to 1. Column $\mathbf{B}^{-1}\mathbf{A}_j, j \notin B$, is the negative of the basic direction subvector $\mathbf{d}_B^j = -\mathbf{B}^{-1}\mathbf{A}_j$. The first entry on the top row or the *0th row*, i.e. $-\mathbf{c}_B^T \mathbf{x}_B$, is the negative of the current cost. The remaining entries in the 0th row are the reduced costs $\overline{c}_1, ..., \overline{c}_N$.

Suppose that $\overline{c}_j < 0$ and that column $\mathbf{A}_j$ is to enter the basis. Define $m^* \in \{1, ..., M\}$ such that column $\mathbf{A}_{B(m^*)}$ leaves the basis. Let $\mathbf{B}'$ denote the new basis, i.e.

$$\mathbf{B}' = [\mathbf{A}_{B(1)} \cdots \mathbf{A}_{B(m^*-1)} \mathbf{A}_j \mathbf{A}_{B(m^*+1)} \cdots \mathbf{A}_{B(M)}].$$

In addition,

$$\mathbf{B}^{-1}\mathbf{B}' = \begin{bmatrix} & -d_{B(1)}^j & \\ \mathbf{e}_1 ... \mathbf{e}_{m^*-1} & -d_{B(m^*)}^j & \mathbf{e}_{m^*+1} ... \mathbf{e}_M \\ & -d_{B(M)}^j & \end{bmatrix}$$

where $\mathbf{e}_i$ is the unit vector with the $i$th component equal to 1.

Let $\mathbf{Q}$ be the matrix such that $\mathbf{Q}\mathbf{B}^{-1}\mathbf{B}' = \mathbf{I}$ or equivalently $\mathbf{Q}\mathbf{B}^{-1} = \mathbf{B}'^{-1}$, where $\mathbf{I}$ is the $M \times M$ identity matrix. Note that transforming $\mathbf{B}^{-1}\mathbf{B}'$ into $\mathbf{I}$ can be done through performing a sequence of *elementary row operations*, which refers to adding a constant multiple of one row to the same or another row.

Notice that the table entries $\mathbf{B}'^{-1}\mathbf{b}$ and $\mathbf{B}'^{-1}\mathbf{A}$ for the next iteration can be obtained by multiplying the current entries $\mathbf{B}^{-1}\mathbf{b}$ and $\mathbf{B}^{-1}\mathbf{A}$ on the left by $\mathbf{Q}$. From the above discussion, multiplication by $\mathbf{Q}$ can be achieved by performing elementary row operations that turn column $\mathbf{B}^{-1}\mathbf{A}_j$ in the table into $\mathbf{e}_{m^*}$. The table entry in the $m^*$th row and in the $j$th column is called the *pivot*. Note that the pivot must be positive.

It remains to update the 0th row of the table. The 0th row can be updated by adding a constant multiple of the $m^*$th row to make $\bar{c}_j = 0$. The justification of this step is given next.

Recall that the 0th row of the table is equal to $[0 \ \mathbf{c}^T] - \mathbf{c}_B^T \mathbf{B}^{-1}[\mathbf{b} \ \mathbf{A}]$. Since the $m^*$th row of the table is the multiplication of the $m^*$th row of $\mathbf{B}^{-1}$ and $[\mathbf{b} \ \mathbf{A}]$, it is a linear combination of $[\mathbf{b} \ \mathbf{A}]$. After adding a constant multiple of the $m^*$th row, the 0th row is transformed into $[0 \ \mathbf{c}^T] - \mathbf{p}^T[\mathbf{b} \ \mathbf{A}]$, where $\mathbf{p}$ is such that $\bar{c}_j = c_j - \mathbf{p}^T \mathbf{A}_j = 0$.

Since $\mathbf{B}^{-1}\mathbf{A}_{B(i)} = \mathbf{e}_i$, in the $m^*$th row of the table, the entries in columns $i$, $i \in \mathcal{B} - \{B(m^*)\}$, are zero. This means that adding the multiple of the $m^*$th row of the table to the 0th row, yielding $[0 \ \mathbf{c}^T] - \mathbf{p}^T[\mathbf{b} \ \mathbf{A}]$, will leave $\bar{c}_i = 0$ for $i \in \mathcal{B} - \{B(m^*)\}$. Together with $\bar{c}_j = 0$, the new basis satisfies $\mathbf{c}_B^T - \mathbf{p}^T \mathbf{B}' = \mathbf{0}$ or equivalently $\mathbf{p}^T = \mathbf{c}_B^T \mathbf{B}'^{-1}$. It follows that, for the new basis, the 0th row of the table is equal to $[0 \ \mathbf{c}^T] - \mathbf{c}_B^T \mathbf{B}'^{-1}[\mathbf{b} \ \mathbf{A}]$ as desired.

The following example shows the operations of the full tableau implementation.

**Example 3.3:** Consider the following linear optimization problem.

$$\text{minimize} - 2x_1 - x_2$$

$$\text{subject to} \begin{bmatrix} 1 & 0 & 1 & 0 & 0 \\ 0 & 1 & 0 & 1 & 0 \\ 1 & 1 & 0 & 0 & 1 \end{bmatrix} \begin{bmatrix} x_1 \\ x_2 \\ x_3 \\ x_4 \\ x_5 \end{bmatrix} = \begin{bmatrix} 3 \\ 2 \\ 4 \end{bmatrix}$$

$$x_1, x_2, x_3, x_4, x_5 \geq 0$$

Starting with the BFS with $\mathbf{B} = [\mathbf{A}_3, \mathbf{A}_4, \mathbf{A}_5]$, i.e. $\mathbf{x} = (0, 0, 3, 2, 4)$, the simplex algorithm proceeds as follows. In each iteration, the pivot is marked with an asterisk.

Iteration 1

| | | $x_1$ | $x_2$ | $x_3$ | $x_4$ | $x_5$ |
|---|---|---|---|---|---|---|
| | 0 | −2 | −1 | 0 | 0 | 0 |
| $x_3 =$ | 3 | 1* | 0 | 1 | 0 | 0 |
| $x_4 =$ | 2 | 0 | 1 | 0 | 1 | 0 |
| $x_5 =$ | 4 | 1 | 1 | 0 | 0 | 1 |

| Iteration 2 | | $x_1$ | $x_2$ | $x_3$ | $x_4$ | $x_5$ |
|---|---|---|---|---|---|---|
| | 6 | 0 | −1 | 2 | 0 | 0 |
| $x_1 =$ | 3 | 1 | 0 | 1 | 0 | 0 |
| $x_4 =$ | 2 | 0 | 1 | 0 | 1 | 0 |
| $x_5 =$ | 1 | 0 | 1* | −1 | 0 | 1 |

| Iteration 3 | | $x_1$ | $x_2$ | $x_3$ | $x_4$ | $x_5$ |
|---|---|---|---|---|---|---|
| | 7 | 0 | 0 | 1 | 0 | 1 |
| $x_1 =$ | 3 | 1 | 0 | 1 | 0 | 0 |
| $x_4 =$ | 1 | 0 | 0 | 1 | 1 | −1 |
| $x_2 =$ | 1 | 0 | 1 | −1 | 0 | 1 |

At this point, all the reduced costs are nonnegative, yielding the optimal solution (3, 1, 0, 1, 0) with the optimal cost −7. Figure 3.5 illustrates how the simplex algorithm moves from one extreme point to another for the feasible set of $(x_1, x_2)$, where all the other variables are viewed as slack variables. In particular, the path taken by the simplex algorithm consists of extreme points (0, 0), (3, 0), and (3, 1).    □

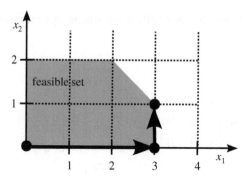

**Figure 3.5**   Illustration of the path taken by the simplex algorithm.

The next example illustrates a potential problem that can occur when some BFSs are degenerate.

**Example 3.4 (from [Papadimitriou and Steiglitz, 1998])**   Consider the following iterations of the simplex algorithm in which pivoting follows the following rules.

1. The nonbasic variable with the most negative $\bar{c}_i$ enters the basis.
2. If $m^*$ is not unique, the basic variable with the smallest index leaves the basis.

Iteration 1

|  |  | $x_1$ | $x_2$ | $x_3$ | $x_4$ | $x_5$ | $x_6$ | $x_7$ |
|---|---|---|---|---|---|---|---|---|
|  | 3 | −3/4 | 20 | −1/2 | 6 | 0 | 0 | 0 |
| $x_5 =$ | 0 | 1/4* | −8 | −1 | 9 | 1 | 0 | 0 |
| $x_6 =$ | 0 | 1/2 | −12 | −1/2 | 3 | 0 | 1 | 0 |
| $x_7 =$ | 1 | 0 | 0 | 1 | 0 | 0 | 0 | 1 |

Iteration 2

|  |  | $x_1$ | $x_2$ | $x_3$ | $x_4$ | $x_5$ | $x_6$ | $x_7$ |
|---|---|---|---|---|---|---|---|---|
|  | 3 | 0 | −4 | −7/2 | 33 | 3 | 0 | 0 |
| $x_1 =$ | 0 | 1 | −32 | −4 | 36 | 4 | 0 | 0 |
| $x_6 =$ | 0 | 0 | 4* | 3/2 | −15 | −2 | 1 | 0 |
| $x_7 =$ | 1 | 0 | 0 | 1 | 0 | 0 | 0 | 1 |

Iteration 3

|  |  | $x_1$ | $x_2$ | $x_3$ | $x_4$ | $x_5$ | $x_6$ | $x_7$ |
|---|---|---|---|---|---|---|---|---|
|  | 3 | 0 | 0 | −2 | 18 | 1 | 1 | 0 |
| $x_1 =$ | 0 | 1 | 0 | 8* | −84 | −12 | 8 | 0 |
| $x_2 =$ | 0 | 0 | 1 | 3/8 | −15/4 | −1/2 | 1/4 | 0 |
| $x_7 =$ | 1 | 0 | 0 | 1 | 0 | 0 | 0 | 1 |

Iteration 4

|  |  | $x_1$ | $x_2$ | $x_3$ | $x_4$ | $x_5$ | $x_6$ | $x_7$ |
|---|---|---|---|---|---|---|---|---|
|  | 3 | 1/4 | 0 | 0 | −3 | −2 | 3 | 0 |
| $x_3 =$ | 0 | 1/8 | 0 | 1 | −21/2 | −3/2 | 1 | 0 |
| $x_2 =$ | 0 | −3/64 | 1 | 0 | 3/16* | 1/16 | −1/8 | 0 |
| $x_7 =$ | 1 | −1/8 | 0 | 0 | 21/2 | 3/2 | −1 | 1 |

Iteration 5

|  |  | $x_1$ | $x_2$ | $x_3$ | $x_4$ | $x_5$ | $x_6$ | $x_7$ |
|---|---|---|---|---|---|---|---|---|
|  | 3 | −1/2 | 16 | 0 | 0 | −1 | 1 | 0 |
| $x_3 =$ | 0 | −5/2 | 56 | 1 | 0 | 2* | −6 | 0 |
| $x_4 =$ | 0 | −1/4 | 16/3 | 0 | 1 | 1/3 | −2/3 | 0 |
| $x_7 =$ | 1 | 5/2 | −56 | 0 | 0 | −2 | 6 | 1 |

Iteration 6

|  |  | $x_1$ | $x_2$ | $x_3$ | $x_4$ | $x_5$ | $x_6$ | $x_7$ |
|---|---|---|---|---|---|---|---|---|
|  | 3 | −7/4 | 44 | 1/2 | 0 | 0 | −2 | 0 |
| $x_5 =$ | 0 | −5/4 | 28 | 1/2 | 0 | 1 | −3 | 0 |
| $x_4 =$ | 0 | 1/6 | −4 | −1/6 | 1 | 0 | 1/3* | 0 |
| $x_7 =$ | 1 | 0 | 0 | 1 | 0 | 0 | 0 | 1 |

Iteration 7

| | | $x_1$ | $x_2$ | $x_3$ | $x_4$ | $x_5$ | $x_6$ | $x_7$ |
|---|---|---|---|---|---|---|---|---|
| | 3 | $-3/4$ | 20 | $-1/2$ | 6 | 0 | 0 | 0 |
| $x_5 =$ | 0 | $1/4^*$ | $-8$ | $-1$ | 9 | 1 | 0 | 0 |
| $x_6 =$ | 0 | $1/2$ | $-12$ | $-1/2$ | 3 | 0 | 1 | 0 |
| $x_7 =$ | 1 | 0 | 0 | 1 | 0 | 0 | 0 | 1 |

Note that the simplex algorithm is back to the initial BFS at iteration 7.                                                                                 □

It is helpful to note that, in example 3.4, the involved BFSs are degenerate. Since each pivot has $\theta^* = 0$, pivoting changes the basic columns without moving away from the previous solution point. When the simplex algorithm continues changing among the same set of basic columns forever, it is said to encounter *cycling*.

One simple method, called the *Bland rule* or the *smallest subscript pivoting rule*, can help avoid cycling. Pivoting based on the Bland rule is described below.

1. Among the nonbasic variable with negative $\bar{c}_i$'s, choose the variable with the smallest index to enter the basis.

2. If $m^*$ is not unique, the basic variable with the smallest index leaves the basis.

The proof that the Bland rule can avoid cycling can be found in [Papadimitriou and Steiglitz, 1998]. The next example shows the simplex iterations according to the Bland rule.

**Example 3.5:**   Assume the same problem as in example 3.4 but with the Bland rule for pivoting. The first four iterations are the same as before. Below are the remaining iterations according to the Bland rule.

Iteration 5

| | | $x_1$ | $x_2$ | $x_3$ | $x_4$ | $x_5$ | $x_6$ | $x_7$ |
|---|---|---|---|---|---|---|---|---|
| | 3 | $-1/2$ | 16 | 0 | 0 | $-1$ | 1 | 0 |
| $x_3 =$ | 0 | $-5/2$ | 56 | 1 | 0 | 2 | $-6$ | 0 |
| $x_4 =$ | 0 | $-1/4$ | $16/3$ | 0 | 1 | $1/3$ | $-2/3$ | 0 |
| $x_7 =$ | 1 | $5/2^*$ | $-56$ | 0 | 0 | $-2$ | 6 | 1 |

Iteration 6

| | | $x_1$ | $x_2$ | $x_3$ | $x_4$ | $x_5$ | $x_6$ | $x_7$ |
|---|---|---|---|---|---|---|---|---|
| | $16/5$ | 0 | $24/5$ | 0 | 0 | $-7/5$ | $11/5$ | $1/5$ |
| $x_3 =$ | 1 | 0 | 0 | 1 | 0 | 0 | 0 | 1 |
| $x_4 =$ | $1/10$ | 0 | $-4/15$ | 0 | 1 | $2/15^*$ | $-1/15$ | $1/10$ |
| $x_1 =$ | $2/5$ | 1 | $-112/5$ | 0 | 0 | $-4/5$ | $12/5$ | $2/5$ |

| Iteration 7 | | $x_1$ | $x_2$ | $x_3$ | $x_4$ | $x_5$ | $x_6$ | $x_7$ |
|---|---|---|---|---|---|---|---|---|
| | 17/4 | 0 | 2 | 0 | 21/2 | 0 | 3/2 | 5/4 |
| $x_3 =$ | 1 | 0 | 0 | 1 | 0 | 0 | 0 | 1 |
| $x_5 =$ | 3/4 | 0 | -2 | 0 | 15/2 | 1 | -1/2 | 3/4 |
| $x_1 =$ | 1 | 1 | -24 | 0 | 6 | 0 | 2 | 1 |

Note that the simplex iterations eventually yield an optimal BFS whose reduced costs are nonnegative. It follows that the resultant optimal solution is (1, 0, 1, 0, 3/4, 0, 0) and the optimal cost is $-17/4$.    □

## Starting the Simplex Algorithm

The discussion on the simplex algorithm up to this point assumes the availability of an initial BFS. This section describes how to find an initial BFS which may not be directly obtained by inspection.

Without loss of generality, assume that $\mathbf{b} \geq \mathbf{0}$. (This condition can be achieved by multiplying some constraints by $-1$ if necessary.) To find an initial BFS of the standard-form problem in (3.4), *M artificial variables*, denoted by $y_1, \ldots, y_M$, are introduced to form the *auxiliary problem*

$$\text{minimize } y_1 + \cdots + y_M$$

$$\text{subject to } \mathbf{Ax} + \mathbf{y} = \mathbf{b}$$

$$\mathbf{x} \geq \mathbf{0}, \mathbf{y} \geq \mathbf{0} \tag{3.7}$$

where $\mathbf{y} = (y_1, \ldots, y_M)$. In what follows, let $\mathbf{f}$ be the cost vector for the auxiliary problem.

Finding a BFS for problem 3.7 is easy. One BFS is $\mathbf{x} = \mathbf{0}$ and $\mathbf{y} = \mathbf{b}$, with $\mathbf{y}$ containing $M$ basic variables. Note that the corresponding basis is the identity matrix. It is clear that the optimal cost of problem 3.7 is at least zero. In addition, if $\mathbf{x}$ is a feasible solution for the original problem, i.e. $\mathbf{Ax} = \mathbf{b}$ and $\mathbf{x} \geq \mathbf{0}$, then $\mathbf{x}$ together with $\mathbf{y} = \mathbf{0}$ is optimal for problem 3.7. Therefore, if the auxiliary problem has a nonzero optimal cost, the original problem is infeasible.

Otherwise, the optimal solution of the auxiliary problem contains a feasible solution $\mathbf{x}$. If $\mathbf{x}$ contains all basic variables, it is a BFS to the original problem. If $\mathbf{x}$ does not contain all basic variables, at least one $y_i$ is basic and is equal to 0. This basic variable $y_i$ can be driven out of the basis by setting $\theta^* = 0$ and choosing a nonzero pivot in the row of $y_i$ and in the column of some nonbasic variable $x_j$, regardless of whether or not the pivot is positive and whether or not the reduced cost $\bar{c}_j$ is negative.[4] Repeating the process for each basic variable $y_i$, a BFS with all the basic variables in $\mathbf{x}$ is eventually obtained.

**Example 3.6:**   Consider the following problem.

$$\text{minimize } x_1 + 2x_2 + 3x_3$$

$$\text{subject to } \begin{bmatrix} 1 & 2 & 1 \\ 1 & 1 & 0 \end{bmatrix} \begin{bmatrix} x_1 \\ x_2 \\ x_3 \end{bmatrix} = \begin{bmatrix} 1 \\ 1 \end{bmatrix}$$

$$x_1, x_2, x_3 \geq 0$$

The corresponding auxiliary problem is given below.

$$\text{minimize } y_1 + y_2$$

$$\text{subject to } \begin{bmatrix} 1 & 2 & 1 & 1 & 0 \\ 1 & 1 & 0 & 0 & 1 \end{bmatrix} \begin{bmatrix} x_1 \\ x_2 \\ x_3 \\ y_1 \\ y_2 \end{bmatrix} = \begin{bmatrix} 1 \\ 1 \end{bmatrix}$$

$$x_1, x_2, x_3, y_1, y_2 \geq 0$$

The initial BFS for the auxiliary problem is $\mathbf{x} = (0, 0, 0)$ and $\mathbf{y} = (1, 1)$. Below are the simplex iterations leading to an optimal solution for the auxiliary problem. Note that the initial reduced cost $\bar{c}_i$ is $- \mathbf{f}_B^T \mathbf{A}_i$, which is the negative of the sum of the components of $\mathbf{A}_i$.

---

[4]Such a pivot must exist. Otherwise, it means that, in the row of $y_i$, $\mathbf{B}^{-1}\mathbf{A}$ is equal to zero. This zero row of $\mathbf{B}^{-1}\mathbf{A}$ implies that one row of $\mathbf{A}$ is a linear combination of the other rows, which contradicts the assumption that $\mathbf{A}$ has full rank.

Iteration 1

|  | $x_1$ | $x_2$ | $x_3$ | $y_1$ | $y_2$ |
|---|---|---|---|---|---|
| $-2$ | $-2$ | $-3$ | $-1$ | $0$ | $0$ |
| $y_1 =$   $1$ | $1^*$ | $2$ | $1$ | $1$ | $0$ |
| $y_2 =$   $1$ | $1$ | $1$ | $0$ | $0$ | $1$ |

Iteration 2

|  | $x_1$ | $x_2$ | $x_3$ | $y_1$ | $y_2$ |
|---|---|---|---|---|---|
| $0$ | $0$ | $1$ | $1$ | $2$ | $0$ |
| $x_1 =$   $1$ | $1$ | $2$ | $1$ | $1$ | $0$ |
| $y_2 =$   $0$ | $0$ | $-1^*$ | $-1$ | $-1$ | $1$ |

At this point, $\mathbf{x} = (1, 0, 0)$ together with $\mathbf{y} = (0, 0)$ is obtained as an optimal solution. However, since $y_2$ is basic, it must be driven out of the basis. One possibility is to let $x_2$ enter the basis, as shown by the pivot in iteration 2, and proceed.

Iteration 3

|  | $x_1$ | $x_2$ | $x_3$ | $y_1$ | $y_2$ |
|---|---|---|---|---|---|
| $0$ | $0$ | $0$ | $0$ | $1$ | $1$ |
| $x_1 =$   $1$ | $1$ | $0$ | $-1$ | $-1$ | $2$ |
| $x_2 =$   $0$ | $0$ | $1$ | $1$ | $1$ | $-1$ |

At this point, $x_1$ and $x_2$ are basic variables, yielding a BFS to the original problem. □

Solving the auxiliary problem to get an initial BFS is referred to as *phase 1* of the simplex algorithm. Continuing with the original problem is referred to as *phase 2*. In summary, the two-phase implementation of the simplex algorithm is as follows.

## Two-phase Simplex Algorithm

### Phase 1

1. Make $\mathbf{b} \geq \mathbf{0}$ by multiplying some constraints with $-1$ if necessary.

2. Form the auxiliary problem by introducing artificial variables $\mathbf{y} = (y_1, ..., y_M)$. Use $\mathbf{x} = \mathbf{0}$ and $\mathbf{y} = \mathbf{b}$ as the initial BFS to solve the auxiliary problem according to the simplex iterations.

3. If the optimal cost is positive, the overall algorithm terminates by concluding that the original problem is infeasible.

4. If the optimal cost is zero, then obtain the initial BFS for the original problem by driving artificial variables $y_i$'s out of the basis if necessary.

**Phase 2**

1. Compute the reduced costs for all nonbasic variables. If all the reduced costs are nonnegative, the algorithm terminates by concluding that the current BFS is optimal.

2. If some reduced cost is negative, select a nonbasic variable $x_j$ with $\overline{c}_j < 0$ to enter the basis. If $\mathbf{d}^j \geq \mathbf{0}$ (yielding $\theta^* = \infty$), the algorithm terminates by concluding that the optimal cost is $-\infty$.

3. If $\theta^*$ is finite, move to another BFS and repeat step 1.

Note that there are three possible terminations. First, the problem is infeasible (termination in phase 1). Second, the optimal cost is $-\infty$ (termination in step 2 of phase 2). Finally, there is an optimal solution (termination in step 1 of phase 2).

## 3.4   Duality for Linear Optimization

Since linear optimization is a special case of convex optimization, duality theory for convex optimization can be applied to linear optimization. In particular, consider again the *primal problem* in the standard form in (3.4).

$$\text{minimize } \mathbf{c}^T\mathbf{x}$$

$$\text{subject to } \mathbf{A}\mathbf{x} = \mathbf{b}$$

$$\mathbf{x} \geq \mathbf{0}$$

Let $\boldsymbol{\lambda} = (\lambda_1, \ldots, \lambda_N)$ and $\boldsymbol{\mu} = (\mu_1, \ldots, \mu_M)$. The Lagrangian for the primal problem is[5]

$$\Lambda(\mathbf{x}, \boldsymbol{\lambda}, \boldsymbol{\mu}) = \mathbf{c}^T\mathbf{x} - \boldsymbol{\lambda}^T\mathbf{x} + \boldsymbol{\mu}^T(\mathbf{b} - \mathbf{A}\mathbf{x}). \tag{3.8}$$

The dual function is

---

[5]In the literature on linear optimization, it is more common to write $\mathbf{h}(\mathbf{x}) = \mathbf{b} - \mathbf{A}\mathbf{x}$ as the equality constraint function. For $\mathbf{h}(\mathbf{x}) = \mathbf{A}\mathbf{x} - \mathbf{b}$, the dual problem will have a different form.

$$q(\lambda, \mu) = \inf_{\mathbf{x} \in \mathbb{R}^N} \Lambda(\mathbf{x}, \lambda, \mu)$$

$$= \inf_{\mathbf{x} \in \mathbb{R}^N} \mathbf{x}^T(\mathbf{c} - \lambda - \mathbf{A}^T\mu) + \mathbf{b}^T\mu = \begin{cases} -\infty, & \mathbf{A}^T\mu + \lambda \neq \mathbf{c} \\ \mathbf{b}^T\mu, & \mathbf{A}^T\mu + \lambda = \mathbf{c} \end{cases} \qquad (3.9)$$

From (3.9), the domain set of the dual function is $\{(\lambda, \mu) \in \mathbb{R}^N \times \mathbb{R}^M |$ $\mathbf{A}^T\mu + \lambda = \mathbf{c}$. Consequently, the *dual problem* is as follows.

> maximize $\mathbf{b}^T\mu$
>
> subject to $\mathbf{A}^T\mu + \lambda = \mathbf{c}$
>
> $\qquad\qquad \lambda \geq 0$ $\qquad\qquad\qquad\qquad (3.10)$

By viewing $\lambda$ in (3.10) as slack variables, the dual problem can also be written as

> maximize $\mathbf{b}^T\mu$
>
> subject to $\mathbf{A}^T\mu \leq \mathbf{c}$ $\qquad\qquad\qquad (3.11)$

From either (3.10) or (3.11), it can be seen that the dual problem of a linear optimization problem is another linear optimization problem. The KKT conditions for the primal-dual optimal solution pair can be applied as follows. From (3.8), Lagrangian optimality yields

$$0 = \frac{\partial \lambda(\mathbf{x}^*, \lambda^*, \mu^*)}{\partial \mathbf{x}} = \mathbf{c} - \lambda^* - \mathbf{A}^T\mu^* \Rightarrow \mathbf{A}^T\mu^* + \lambda^* = \mathbf{c}, \qquad (3.12)$$

which is equivalent to one constraint in the dual problem. Complimentary slackness requires that $\lambda_i^* x_i^* = 0$, $i \in \{1, ..., N\}$. From (3.12), either $x_i^* = 0$ or $\mu^{*T}\mathbf{A}_i = c_i$. Equivalently,

$$x_i^*(c_i - \mu^{*T}\mathbf{A}_i) = 0, \ i \in \{1, ..., N\}. \qquad (3.13)$$

From the weak duality theory, any dual feasible solution yields a lower bound on the primal optimal cost. More specifically, let $(\lambda, \mu)$ be dual feasible, and $\mathbf{x}^*$ be a primal optimal solution. It follows that

$$\mathbf{b}^T\mu \leq \mathbf{c}^T\mathbf{x}^*. \qquad (3.14)$$

Let $f^*$ and $q^*$ be the primal and dual optimal costs respectively. From weak duality, if $q^* = \infty$, the primal problem is infeasible. On the other hand, if $f^* = -\infty$, the dual problem is infeasible. The following theorem states that, for a linear optimization problem with an optimal solution, strong duality always hold, i.e. there is no duality gap.

**Theorem 3.7 (Strong duality for linear optimization)**   If the linear optimization problem in the standard form in (3.4) has an optimal solution, so does its dual problem in (3.10) or (3.11). In addition, their optimal costs are equal, i.e. $f^* = q^*$.

**Proof**   Consider a primal optimal solution $\mathbf{x}^*$ obtained from the simplex algorithm. Let $\mathbf{B}$ be the basis matrix for $\mathbf{x}^*$. Let $\mathbf{x}_B^* = \mathbf{B}^{-1}\mathbf{b}$ be the vector of basic variables. From the optimality condition, the reduced cost vector of $\mathbf{x}^*$ is nonnegative, i.e. $\mathbf{c}^T - \mathbf{c}_B\mathbf{B}^{-1}\mathbf{A} \geq \mathbf{0}$, where $\mathbf{c}_B$ is the cost vector for basic variables.

Define a vector $\boldsymbol{\mu}^*$ such that $\boldsymbol{\mu}^{*T} = \mathbf{c}_B^T\mathbf{B}^{-1}$. Note that $\mathbf{c}^T - \boldsymbol{\mu}^{*T}\mathbf{A} \geq \mathbf{0}$ or equivalently $\mathbf{A}^T\boldsymbol{\mu}^* \leq \mathbf{c}$. Thus, $\boldsymbol{\mu}^*$ is a feasible solution to problem 3.11. In addition, the cost of $\boldsymbol{\mu}^*$ is equal to

$$\boldsymbol{\mu}^{*T}\mathbf{b} = \mathbf{c}_B^T\mathbf{B}^{-1}\mathbf{b} = \mathbf{c}_B^T\mathbf{x}_B^* = \mathbf{c}^T\mathbf{x}^*,$$

which from weak duality implies that $\boldsymbol{\mu}^*$ is a dual optimal solution. Finally, the above equality implies that $f^* = q^*$.    □

## †Dual Simplex Algorithm

The proof of theorem 3.7 shows that the simplex algorithm yields a primal optimal solution $\mathbf{x}^*$ as well as a dual optimal solution $\boldsymbol{\mu}^{*T} = \mathbf{c}_B^T\mathbf{B}^{-1}$. Notice that, throughout the simplex iterations, the primal feasibility, i.e. $\mathbf{A}\mathbf{x} = \mathbf{b}$ and $\mathbf{x} \geq \mathbf{0}$, is maintained. Let $\boldsymbol{\mu}^T = \mathbf{c}_B^T\mathbf{B}^{-1}$ contain the dual variables. The dual feasibility condition $\mathbf{A}^T\boldsymbol{\mu} \leq \mathbf{c}$ in (3.11) is obtained after an optimal solution is found since the condition is equivalent to having nonnegative reduced costs, i.e. $\mathbf{c}^T - \mathbf{c}_B^T\mathbf{B}^{-1}\mathbf{A} \geq \mathbf{0}$. In this regards, an algorithm that maintains the primal feasibility is called a *primal algorithm*.

As an alternative, a *dual algorithm* maintains the dual feasibility and obtains the primal feasibility after an optimal solution is found. This alternative yields the *dual simplex algorithm*, which maintains the same table as before and as repeated below.

| $-\mathbf{c}_B^T \mathbf{B}^{-1} \mathbf{b}$ | $\mathbf{c}^T - \mathbf{c}_B^T \mathbf{B}^{-1} \mathbf{A}$ |
|---|---|
| $\mathbf{B}^{-1} \mathbf{b}$ | $\mathbf{B}^{-1} \mathbf{A}$ |

Note that the top left corner $-\mathbf{c}_B^T \mathbf{B}^{-1} \mathbf{b}$ is also equal to the negative of the dual cost since $\boldsymbol{\mu}^T = \mathbf{c}_B^T \mathbf{B}^{-1}$. Since $\mathbf{A}^T \boldsymbol{\mu} \leq \mathbf{c}$ is equivalent to $\mathbf{c}^T - \mathbf{c}_B^T \mathbf{B}^{-1} \mathbf{A} \geq \mathbf{0}$, the top right corner must be nonnegative to maintain the dual feasibility.

While maintaining the dual feasibility, if $\mathbf{B}^{-1} \mathbf{b} \geq \mathbf{0}$, then the primal feasibility is obtained together with nonnegative reduced costs. It follows that the current BFS is an optimal solution. On the other hand, if a basic variable is negative, the dual simplex algorithm performs pivoting as follows.

1. Pick one negative basic variable, say $x_{B(i)}$. Use the row of $x_{B(i)}$, i.e. row $i$ of the table, as a pivoting row.

2. Recall that $\mathbf{d}^j = -\mathbf{B}^{-1} \mathbf{A}_j$. Note that row $i$ of the table contains $-d_i^1, \ldots, -d_i^N$. If $d_i^j \leq 0$ for all $j \in \{1, \ldots, N\}$, the dual cost can be increased indefinitely by subtracting an arbitrarily large positive multiple of row $i$ from row 0. In this case, the corresponding dual optimal cost can be increased to $\infty$, implying the infeasibility of the primal problem. The algorithm then terminates by concluding that the primal problem is infeasible.

3. If $d_N^j > 0$ for some $j \in \{1, \ldots, N\}$, then let $\theta^* = \min_{j \in B, \, d_i^j > 0} (\bar{c}_j / d_i^j)$, $j^* = \arg \min_{j \in B, \, d_i^j > 0} (\bar{c}_j / d_i^j)$, and use column $j^*$ as the pivoting column. Perform pivoting as in the primal algorithm, i.e. so that column $j^*$ of the table contains 0 in row 0 and the unit vector with the $B(i)$th component equal to 1 in rows 1 to $M$.

From the choice of $\theta^*$ and $j^*$, it follows that the reduced costs are always nonnegative. Therefore, dual feasibility is maintained throughout. Similar to the primal algorithm, the Bland rule can be used to avoid cycling.

**Example 3.7 (from [Bertsimas and Tsitsiklis, 1997])** Below is an example of an iteration of the dual simplex algorithm.

| Iteration 1 | | $x_1$ | $x_2$ | $x_3$ | $x_4$ | $x_5$ |
|---|---|---|---|---|---|---|
| | 0 | 2 | 6 | 10 | 0 | 0 |
| $x_4 =$ | 2 | -2 | 4 | 1 | 1 | 0 |
| $x_5 =$ | -1 | 4 | $-2^*$ | -3 | 0 | 1 |

| Iteration 2 | | $x_1$ | $x_2$ | $x_3$ | $x_4$ | $x_5$ |
|---|---|---|---|---|---|---|
| | -3 | 14 | 0 | 1 | 0 | 3 |
| $x_4 =$ | 0 | 6 | 0 | -5 | 1 | 2 |
| $x_2 =$ | 1/2 | -2 | 1 | 3/2 | 0 | -1/2 |

After one iteraton, note that the current solution is optimal.    □

The choice between primal and dual algorithms depends on application scenarios. If the same linear optimization problem must be solved for different values of $c$, the primal algorithm is appropriate since all problem instances can start with the same primal BFS, i.e. changing $c$ does not change the primal BFSs. On the other hand, if the same problem must be solved for different values of $b$, the dual algorithm is more attractive after the first optimal solution is obtained since all subsequent problem instances can start with the same dual feasible solution, i.e. changing $b$ does not change the dual feasible solutions.

## †Farkas Lemma

*Farkas lemma* was developed before duality theory for linear optimization [Bertsimas and Tsitsiklis, 1997]. Nevertheless, duality theory leads to a simple derivation of Farkas lemma, which is worth mentioning since it can be used to derive several important results in applied mathematics.

**Theorem 3.8 (Farkas lemma)**   Let $b$, $\mu \in \mathbb{R}^M$ and $A \in \mathbb{R}^M \times \mathbb{R}^N$ (i.e. $A$ is an $M \times N$ matrix.) Then, exactly one of the two statements is true.

1. There exists an $x \geq 0$ such that $Ax = b$.

2. There exists a $\mu$ such that $A^T\mu \geq 0$ and $b^T\mu < 0$.

**Proof**   First, argue that if statement 1 is true, then statement 2 is false. To see this, consider an $x \geq 0$ with $Ax = b$. Since $b^T\mu = x^TA^T\mu$, having $A^T\mu \geq 0$ together with $x \geq 0$ yields $b^T\mu \geq 0$, implying that statement 2 is false.

It remains to show that if statement 1 is false, then statement 2 must be true. Consider the following primal-dual pair of linear optimization problems.

$$\text{minimize } \mathbf{0}^T\mathbf{x} \qquad\qquad \text{minimize } \mathbf{b}^T\boldsymbol{\mu}$$

$$\text{subjuct to } -\mathbf{A}\mathbf{x} = -\mathbf{b} \quad\Leftrightarrow\quad \text{subject to } \mathbf{A}^T\boldsymbol{\mu} \geq \mathbf{0}$$

$$\mathbf{x} \geq 0$$

If statement 1 is false, then the primal problem is infeasible, implying that the dual problem does not have an optimal solution. Therefore, the dual problem is either infeasible or has the optimal cost equal to $-\infty$. Since $\boldsymbol{\mu} = \mathbf{0}$ is dual feasible, it follows that the dual optimal cost is $-\infty$, implying that statement 2 is true. □

**Example 3.8 (Using Farkas lemma)** As an example usage of Farkas lemma, consider proving that, for any finite-state Markov chain, there exists a set of steady-state probabilities [Bertsimas and Tsitsiklis, 1997].

A *finite-state Markov chain* consists of a finite set of states indexed from 1 to $N$, where $N$ is the number of states. At any time, the system is in one of the $N$ states. When a transition from state $i$ occurs, the next state is state $j$ (with $j = i$ possible) with probability $p_{ij}$. Note that $\sum_{j=1}^{N} p_{ij} = 1$ for each state $i$. Let $N \times N$ matrix $\mathbf{P}$ contain all the transition probabilities, i.e. with $p_{ij}$ as the entry in row $i$ and column $j$.

An $N \times N$ matrix $\mathbf{A}$ is *stochastic* if all entries are nonnegative and the sum of entries in each row is equal to 1. Note that the transition probability matrix $\mathbf{P}$ is stochastic since $\sum_{j=1}^{N} p_{ij} = 1$ for each $i$.

In what follows, Farkas lemma will be used to show that, for any stochastic matrix $\mathbf{P}$, there exists a vector $\mathbf{x} \geq \mathbf{0}$ such that $\mathbf{x}^T\mathbf{P} = \mathbf{x}^T$ and $\mathbf{x}^T\mathbf{e} = 1$, where $\mathbf{e}$ is the $N \times 1$ vector with all components equal to 1. Note that $\mathbf{x}$ can be viewed as containing the steady-state probabilities.

The constraints $\mathbf{x}^T\mathbf{P} = \mathbf{x}^T$, $\mathbf{x}^T\mathbf{e} = 1$, and $\mathbf{x} \geq \mathbf{0}$ can be expressed more compactly as

$$\begin{bmatrix} \mathbf{P}^T - \mathbf{I} \\ \hline \mathbf{e}^T \end{bmatrix} \mathbf{x} = \begin{bmatrix} \mathbf{0} \\ 1 \end{bmatrix} \text{ and } \mathbf{x} \geq \mathbf{0},$$

where $\begin{bmatrix} \mathbf{A} \\ \hline \mathbf{B} \end{bmatrix}$ denotes the matrix that contains the rows of $\mathbf{A}$ and $\mathbf{B}$.[6]

---

[6]For example, if $\mathbf{A} = [1\ 2]$ and $\mathbf{B} = [3\ 4]$, then $\begin{bmatrix} \mathbf{A} \\ \hline \mathbf{B} \end{bmatrix} = \begin{bmatrix} 1 & 2 \\ 3 & 4 \end{bmatrix}$.

Similarly, let $[\mathbf{A}|\mathbf{B}]$ denote the matrix that contains the columns of $\mathbf{A}$ and $\mathbf{B}$.[7]

Consider statement 2 of Farkas lemma. It is next argued by contradiction that there is no $\boldsymbol{\mu}$ such that

$$[\mathbf{P} - \mathbf{I}|\mathbf{e}]\boldsymbol{\mu} \geq \mathbf{0} \text{ and } \mu_{N+1} < 0.$$

Suppose such a $\boldsymbol{\mu}$ exists. Let $i^* = \arg\max_{i \in \{1,...,N\}} \mu_i$ and $\mu_{\max} = \mu_{i^*}$. Then, row $i^*$ of the constraint $[\mathbf{P} - \mathbf{I}|\mathbf{e}]\boldsymbol{\mu} \geq \mathbf{0}$ can be written as

$$\sum_{j=1}^{N} p_{i^*j}\,\mu_j - \mu_{\max} + \mu_{N+1} \geq 0.$$

Since $\mu_j \leq \mu_{\max}$ for all $j \in \{1, ..., N\}$, it follows that $\sum_{j=1}^{N} p_{i^*j}\,\mu_j \leq$ $\mu_{\max} \sum_{j=1}^{N} p_{i^*j} = \mu_{\max}$, where the last equality follows from the fact that $\mathbf{P}$ is stochastic. From $\mu_{N+1} < 0$,

$$\sum_{j=1}^{N} p_{i^*j}\,\mu_j - \mu_{\max} + \mu_{N+1} \leq \mu_{\max} - \mu_{\max} + \mu_{N+1} = \mu_{N+1} < 0,$$

yielding a contradiction.

Since there is no $\boldsymbol{\mu}$ such that $[\mathbf{P} - \mathbf{I}|\mathbf{e}]\boldsymbol{\mu} \geq \mathbf{0}$ and $\mu_{N+1} < 0$, Farkas lemma implies that there is an $\mathbf{x} \geq \mathbf{0}$ such that $\begin{bmatrix} \mathbf{P}^{\mathrm{T}} - \mathbf{I} \\ \mathbf{e}^{\mathrm{T}} \end{bmatrix} \mathbf{x} = \begin{bmatrix} \mathbf{0} \\ 1 \end{bmatrix}.$   □

## 3.5   Application: Minimum Cost Routing

Consider a *static* routing problem whose goal is to find paths to support a given traffic demand in a network. Assume that transmission link capacities are given. In addition, assume that the traffic stream of a connection can be split onto multiple paths from the source to the destination.

---

[7]For example, if $\mathbf{A} = \begin{bmatrix} 1 \\ 2 \end{bmatrix}$ and $\mathbf{B} = \begin{bmatrix} 3 \\ 4 \end{bmatrix}$, then $[\mathbf{A}|\mathbf{B}] = \begin{bmatrix} 1 & 3 \\ 2 & 4 \end{bmatrix}.$

Let $\mathcal{L}$ denote the set of all directed links. Let $c_l$, $l \in \mathcal{L}$, denote the capacity of link $l$. Let $f_l$ be the total flow (summed over all source-destination (s-d) pairs) assigned on link $l$. One common objective of routing is to minimizing the total cost of the form [Bertsimas and Tsitsiklis, 1997]

$$\sum_{l \in L} \alpha_l f_l, \tag{3.15}$$

where $\alpha_l$ is the cost per unit capacity of using link $l$, e.g. $\alpha_l$ can reflect the distance (in km) of $l$. Another common objective of routing is to minimize the maximum link load or link utilization, i.e.

$$\max_{l \in \mathcal{L}} (f_l / c_l). \tag{3.16}$$

Based on the objective in (3.15), the static routing problem can be formulated as a linear optimization problem as follows.

## Given information

- $\mathcal{L}$: set of directed links
- $c_l$: transmission capacity (in bps) of link $l$
- $\alpha_l$: cost per unit capacity in using link $l$
- $\mathcal{S}$: set of s-d pairs (with nonzero traffic)
- $t^s$: traffic demand (in bps) for s-d pair $s$
- $\mathcal{P}^s$: set of candidate paths for s-d pair $s$
- $\mathcal{P}_l$: set of candidate paths that use link $l$

## Variables

- $x^p$: traffic flow (in bps) on path $p$ (Note that path $p$ implies a particular s-d pair.)

## Constraints

- Satisfaction of traffic demands

$$\forall s \in \mathcal{S}, \quad \sum_{p \in \mathcal{P}^s} x^p = t^s$$

- Link capacity constraint

$$\forall l \in \mathcal{L}, \sum_{p \in P_l} x^p \leq c_l$$

- Non-negativity of traffic flows

$$\forall s \in \mathcal{S}, \forall p \in \mathcal{P}^s, x^p \geq 0$$

## *Objective*

- Minimize the total cost of using links (Note that $\sum_{p \in P_l} x^p = f_l$ in the above discussion.)

$$\text{minimize} \sum_{l \in L} \alpha_l \left( \sum_{p \in P_l} x^p \right)$$

The overall optimization problem is as follows.

$$
\begin{array}{l}
\text{minimize} \ \sum_{l \in L} \alpha_l \left( \sum_{p \in P_l} x^p \right) \\[2em]
\text{subject to} \ \forall s \in \mathcal{S}, \ \sum_{p \in \mathcal{P}^s} x^p = t^s \\[2em]
\forall l \in \mathcal{L}, \ \sum_{p \in P_l} x^p \leq c_l \\[2em]
\forall s \in \mathcal{S}, \forall p \in \mathcal{P}^s, x^p \geq 0
\end{array}
\tag{3.17}
$$

For the objective in (3.16), the objective function is expressed as

$$\text{minimize} \ \max_{l \in \mathcal{L}} \left( \frac{\sum_{p \in P_l} x^p}{c_l} \right),$$

which can be rewritten using an additional variable $f_{\text{max}}$ to denote the maximum link load and additional constraints as shown below.

$$\text{minimize } f_{\max}$$

$$\text{subject to } \forall l \in \mathcal{L}, f_{\max} \geq \frac{\sum_{p \in P_l} x^p}{c_l}$$

**Example 3.9:** Consider the minimum cost routing problem in (3.17) for the network shown in figure 3.6. Note that there are four nodes and seven directed links. The number on each link indicates the link capacity. Assume that $\alpha_l = 1$ for all $l \in \mathcal{L}$. In addition, assume that there are four s-d pairs with nonzero traffic, as shown below together with their candidate paths.

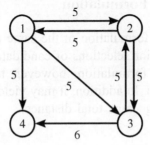

**Figure 3.6** Example network for minimum cost routing of static traffic.

| s-d pair | Demand | Candidate paths |
|----------|--------|-----------------|
| 1-2 | 3 | $1 \to 2$, $1 \to 3 \to 2$ |
| 1-3 | 4 | $1 \to 3$, $1 \to 2 \to 3$ |
| 1-4 | 5 | $1 \to 4$, $1 \to 3 \to 4$ |
| 2-4 | 6 | $2 \to 1 \to 4$, $2 \to 3 \to 4$ |

Note that there are eight variables. Using the `glpk` command in Octave [Octave, online], the following primal optimal solution is obtained with the optimal cost equal to 25.[8]

$$(x^{1 \to 2*}, x^{1 \to 3 \to 2*}, x^{1 \to 3*}, x^{1 \to 2 \to 3*}, x^{1 \to 4*}, x^{1 \to 3 \to 4*}, x^{2 \to 1 \to 4*}, x^{2 \to 3 \to 4*})$$

$$= (3, 0, 4, 0, 4, 1, 1, 5)$$

The Lagrange multipliers for the link capacity constraints are

$$(\lambda^*_{(1, 2)}, \lambda^*_{(2, 1)}, \lambda^*_{(2, 3)}, \lambda^*_{(3, 2)}, \lambda^*_{(1, 3)}, \lambda^*_{(1, 4)}, \lambda^*_{(3, 4)})$$

$$= (0, 0, 1, 0, 0, 1, 0).$$

---

[8]See section C.3 in the appendix for specific Octave commands.

To verify the sensitivity information contained in the Lagrange multipliers. Consider expanding 1 unit of link capacity in two cases.

- Increase $c_{(2, 1)}$ from 5 to 6: The optimal cost is unchanged, i.e. equal to 25. This result is expected since the Lagrange multiplier for the capacity constraint of link (2, 1) is 0.

- Increase $c_{(2, 3)}$ from 5 to 6: The new primal optimal solution is (3, 0, 4, 0, 5, 0, 0, 6), with the optimal cost equal to 24. This result is expected since the Lagrange multiplier for the capacity constraint of link (2,3) is 1.                                    □

## Alternative Problem Formulation

Below is an alternative formulation of the same routing problem which does not assume any prior selections of candidate paths [Bertsimas and Tsitsiklis, 1997]. This formulation, however, tends to contain more variables and constraints. In addition, it may yield paths that may not be desirable in practice, e.g. large total distance.

### *Given information*

- $\mathcal{N}$:    set of nodes
- $\mathcal{L}$:    set of directed links
- $\mathcal{L}_n^{OUT}$: set of links out of node $n$
- $\mathcal{L}_n^{IN}$:  set of links into node $n$
- $c_l$:    transmission capacity (in bps) of link $l$
- $\alpha_l$:    cost per unit capacity in using link $l$
- $\mathcal{S}$:    set of s-d pairs (with nonzero traffic)
- $t^s$:    traffic demand (in bps) for s-d pair $s$

### *Variables*

- $x_l^s$: traffic flow (in bps) on link $l$ for s-d pair $s$

## Constraints

- Flow conservation and satisfaction of traffic demands

$$\forall n \in \mathcal{N}, \forall s \in \mathcal{S}, \sum_{l \in \mathcal{L}_n^{\text{IN}}} x_l^s - \sum_{l \in \mathcal{L}_n^{\text{OUT}}} x_l^s = \begin{cases} -t^s, & n = \text{source of } s \\ t^s, & n = \text{destination of } s \\ 0, & \text{otherwise} \end{cases}$$

- Link capacity constraint

$$\forall l \in \mathcal{L}, \sum_{s \in \mathcal{S}} x_l^s \leq c_l$$

- Non-negativity of traffic flows

$$\forall s \in \mathcal{S}, \forall l \in \mathcal{L}, x_l^s \geq 0$$

## Objective

- Minimize the total cost of using links

$$\text{minimize} \sum_{l \in L} \alpha_l \left( \sum_{s \in \mathcal{S}} x_l^s \right)$$

The overall optimization problem is as follows.

$$\text{minimize} \sum_{l \in L} \alpha_l \left( \sum_{s \in \mathcal{S}} x_l^s \right)$$

subject to $\forall n \in \mathcal{N}, \forall s \in \mathcal{S},$

$$\sum_{l \in \mathcal{L}_n^{\text{IN}}} x_l^s - \sum_{l \in \mathcal{L}_n^{\text{OUT}}} x_l^s = \begin{cases} -t^s, & n = \text{source of } s \\ t^s, & n = \text{destination of } s \\ 0, & \text{otherwise} \end{cases} \quad (3.18)$$

$$\forall l \in \mathcal{L}, \sum_{s \in \mathcal{S}} x_l^s \leq c_l$$

$$\forall s \in S, \forall l \in \mathcal{L}, x_l^s \geq 0$$

For the objective in (3.16), the objective is expressed as

$$\text{minimize} \ \max_{l \in \mathcal{L}} \ \left( \frac{\sum_{s \in S} x_l^s}{c_l} \right),$$

which can be rewritten using an additional variable $f_{\max}$ to denote the maximum link load and additional constraints as shown below.

$$\text{minimize} \ f_{\max}$$

$$\text{subject to} \ \forall l \in \mathcal{L}, f_{\max} \geq \frac{\sum_{s \in S} x_l^s}{c_l}$$

## 3.6 Application: Maximum Lifetime Routing in a WSN

A *wireless sensor network* (*WSN*) consists of a collection of sensor nodes distributed geographically in its deployment field. Several types of sensor nodes exist, e.g. temperature and humidity sensors. In several applications, sensor nodes transmit their data wirelessly to a central processing node once in every fixed time interval. A sensor node may transmit directly to the central node (if it can be reached), or transmit through its neighbors, i.e. multi-hop routing.

Energy consumption is a major concern in a WSN. With less consumption, a sensor node can last longer before any maintainance is needed, e.g. battery change. Since different routings of data incur different amounts of energy consumption, the problem of routing to minimize the energy consumption is of practical interest.

The discussion in this section is based on the framework in [Heinzelman et al., 2002; Chang and Tassiulas, 2004]. For simplicity, it is assumed that most energy consumption by sensor nodes is for transmitting and receiving data packets. In addition, assume that packets have a fixed size. More specifically, the given information and the decision variables are as follows.

## Given information

- $N$: set of nodes (including node 0 which is the central node)
- $N'$: set of sensor nodes (not including node 0)
- $N_i$: set of nodes within the transmission range of node $i$ (not including node $i$)
- $t^k$: traffic rate from sensor node $k$ to the central node
- $E_i$: initial battery energy of sensor node $i$
- $\alpha_{ij}^t$: energy consumption by node $i$ to transmit a packet to node $j$
- $\alpha_i^r$: energy consumption by node $i$ to receive a packet

## Variables

- $f_{ij}^k$: traffic flow from node $i$ to node $j$ for data of sensor node $k$

Given a routing solution, i.e. values of of $f_{ij}^k$, the *lifetime* of node $i$ is computed as

$$L_i = \frac{E_i}{\sum_{k \in N'} \sum_{j \in N_i} \alpha_{ij}^t f_{ij}^k + \sum_{k \in N'} \sum_{j \in N_i} \alpha_i^r f_{ji}^k} \qquad (3.19)$$

The *network lifetime* is defined as the time until one node runs out of battery, i.e.

$$L_{\text{net}} = \min_{i \in N'} L_i. \qquad (3.20)$$

In *maximum lifetime routing* (MLR) proposed in [Chang and Tassiulas, 2004], the objective of routing is to maximize $L_{\text{net}}$. The routing constraints are given below.

## Constraints

- Flow conservation

$$\forall i \in N, \forall k \in N', \quad \sum_{j \in N_i} f_{ji}^k - \sum_{j \in N_i} f_{ji}^k = \begin{cases} -t^k, & i = k \\ t^k, & i = 0 \\ 0, & \text{otherwise} \end{cases} \qquad (3.21)$$

- Non-negativity of flows

$$\forall k \in \mathcal{N}', \; \forall i \in \mathcal{N}, \; \forall j \in \mathcal{N}_i, \; f_{ij}^k \geq 0$$

The overall MLR problem in [Chang and Tassiulas, 2004] is not an LP problem since the cost function in (3.20) is not linear. In particular, the expression of node lifetime $L_i$ in (3.19) is not linear. Based on an observation in [Phyo, 2006], it is convenient to define the *inverse lifetime* of node $i$ as $L_i^{-1} = 1/L_i$. Accordingly, the *inverse network lifetime* is defined as $L_{\text{net}}^{-1} = 1/L_{\text{net}}$. It follows that maximizing $L_{\text{net}} = \min_{i \in \mathcal{N}'} L_i$ is equivalent to minimizing $L_{\text{net}}^{-1} = \max_{i \in \mathcal{N}'} L_i^{-1}$. The LP problem for MLR in WSNs can be expressed as follows.

$$
\begin{aligned}
\text{minimize } \; & L_{\text{net}}^{-1} \\
\text{subject to } \; & \forall i \in \mathcal{N}', \; L_{\text{net}}^{-1} \\[4pt]
& \geq \frac{\sum_{k \in \mathcal{N}'} \sum_{j \in \mathcal{N}_i} \alpha_{ij}^t f_{ij}^k + \sum_{k \in \mathcal{N}'} \sum_{j \in \mathcal{N}_i} \alpha_i^r f_{ji}^k}{E_i} \\[4pt]
& \forall i \in \mathcal{N}, \; \forall k \in \mathcal{N}', \; \sum_{j \in \mathcal{N}_i} f_{ji}^k - \sum_{j \in \mathcal{N}_i} f_{ij}^k \\[4pt]
& = \begin{cases} -t^k, & i = k \\ t^k, & i = 0 \\ 0, & \text{otherwise} \end{cases} \\[4pt]
& \forall k \in \mathcal{N}', \; \forall i \in \mathcal{N}, \; \forall j \in \mathcal{N}_i, \; f_{ij}^k \geq 0
\end{aligned}
\tag{3.22}
$$

Specific values of $\alpha_{ij}^t$ and $\alpha_i^r$ can be obtained from [Heinzelman et al., 2002]. In particular, the coefficients are of the forms

$$\alpha_{ij}^t = P(E_{\text{elec}} + E_{\text{amp}} \, d_{ij}^{\gamma}), \quad \alpha_i^r = P E_{\text{elec}}, \tag{3.23}$$

where $P$ is the packet size (in bit), $E_{\text{elec}}$ is the energy consumption for transmitting or receiving one bit, $E_{\text{amp}}$ is the energy dissipation of the transmit amplifier, $d_{ij}$ is the distance between node $i$ and node $j$, and $\gamma$ is the path loss exponent. In [Heinzelman et al., 2002], the parameter values are $E_{elec}$ = 50 nJ/bit, $E_{amp}$ = 100 pJ/bit/m$^2$, and $\gamma$ = 2.

**Example 3.10:** Consider the MLR problem in (3.22) for the small network shown in figure 3.7. Note that there are four nodes. Assume that all nodes are within transmission ranges of one another, yielding a total of nine links; the three links from node 0 are not considered since they are never used. In addition, assume that $PE_{elec} = PE_{amp} = 1$, $n = 2$, $t^1 = t^2 = t^3 = 1$, and $E_1 = E_2 = E_3 = 10$, where the physical units are omitted in this example for simplicity.

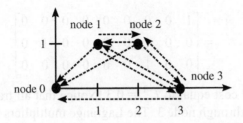

**Figure 3.7** Example network for MLR in a WSN.

Using the `glpk` command in Octave, the obtained optimal routing solution is shown below; the corresponding optimal cost is $L_{net}^{-1*} = 0.633$.[9]

$$
\begin{bmatrix}
x_{10}^{1*} & x_{20}^{1*} & x_{30}^{1*} & x_{12}^{1*} & x_{21}^{1*} & x_{13}^{1*} & x_{31}^{1*} & x_{23}^{1*} & x_{32}^{1*} \\
x_{10}^{2*} & x_{20}^{2*} & x_{30}^{2*} & x_{12}^{2*} & x_{21}^{2*} & x_{13}^{2*} & x_{31}^{2*} & x_{23}^{2*} & x_{32}^{2*} \\
x_{10}^{3*} & x_{20}^{3*} & x_{30}^{3*} & x_{12}^{3*} & x_{21}^{3*} & x_{13}^{3*} & x_{31}^{3*} & x_{23}^{3*} & x_{32}^{3*}
\end{bmatrix}
$$

$$
=
\begin{bmatrix}
1 & 0 & 0 & 0 & 0 & 0 & 0 & 0 & 0 \\
0.833 & 0.167 & 0 & 0 & 0.833 & 0 & 0 & 0 & 0 \\
0 & 0.524 & 0.476 & 0 & 0 & 0 & 0 & 0 & 0.524
\end{bmatrix}
$$

The Lagrange multipliers for the lifetime constraints with $L_{net}^{-1}$ are

$$(\lambda_1^*, \lambda_2^*, \lambda_3^*) = (0.333, 0.333, 0.333),$$

---

[9]See section C.2 in the appendix for specific Octave commands.

which from complimentary slackness implies that nodes 1, 2, and 3 will run out of battery energy at the same time, i.e. they have the same node lifetime.

Suppose $E_3$ is changed from 10 to 100, i.e. node 3 has practically unlimited battery energy. Then, the optimal routing solution becomes

$$
\begin{bmatrix}
x_{10}^{1*} & x_{20}^{1*} & x_{30}^{1*} & x_{12}^{1*} & x_{21}^{1*} & x_{13}^{1*} & x_{31}^{1*} & x_{23}^{1*} & x_{32}^{1*} \\
x_{10}^{2*} & x_{20}^{2*} & x_{30}^{2*} & x_{12}^{2*} & x_{21}^{2*} & x_{13}^{2*} & x_{31}^{2*} & x_{23}^{2*} & x_{32}^{2*} \\
x_{10}^{3*} & x_{20}^{3*} & x_{30}^{3*} & x_{12}^{3*} & x_{21}^{3*} & x_{13}^{3*} & x_{31}^{3*} & x_{23}^{3*} & x_{32}^{3*}
\end{bmatrix}
$$

$$
=
\begin{bmatrix}
1 & 0 & 0 & 0 & 0 & 0 & 0 & 0 & 0 \\
0 & 0 & 1 & 0 & 0 & 0 & 0 & 1 & 0 \\
0 & 0 & 1 & 0 & 0 & 0 & 0 & 0 & 0
\end{bmatrix}
$$

with the optimal cost equal to $L_{\text{net}}^{-1} = 0.3$. Notice that all traffic from node 2 is now routed through node 3. The Lagrange multipliers for the lifetime constraints with $L_{\text{net}}^{-1}$ are

$$(\lambda_1^*, \lambda_2^*, \lambda_3^*) = (0.2, 0.8, 0),$$

which from complimentary slackness implies that nodes 1 and 2 will run out of battery energy at the same time while node 3 may operate longer.                                                                          □

## 3.7   Exercise Problems

**Problem 3.1 (Geometric visualization of linear optimization)**
Consider the following linear optimization problem.

$$\text{minimize } -2x_1 - x_2$$

$$\text{subject to } x_1 \leq 3, \ x_2 \leq 2, \ x_1 + x_2 \leq 4$$

(a) Draw the feasible set for the given problem.

(b) Identify all the extreme points of the feasible set in part (a).

(c) Draw the contour lines for the costs −4 and −6.

(d) By inspection, specify an optimal solution. What is the associated optimal cost? Is the optimal solution unique?

(e) Give an example of cost coefficients for which there are infinitely many optimal solutions.

(f) Give an example of cost coefficients for which there is no optimal solution.

**Problem 3.2 (Computation of BFSs, feasible directions, and reduced costs)** Consider the following linear optimization problem.

$$\text{minimize} - 5x_1 - 3x_2$$

$$\text{subject to } 2x_1 + x_2 \le 8$$

$$x_1 + 2x_2 \le 6$$

$$x_1, x_2 \ge 0$$

(a) Convert the problem to the standard form.

(b) Find all the BFSs in which $x_1$ is *not* a basic variable while $x_2$ is a basic variable.

(c) Find the basic solution in which $x_1$ and $x_2$ are basic variables. Is this basic solution a BFS?

(d) For one BFS in part (a), find the basic direction and the reduced cost for each of the nonbasic variables.

**Problem 3.3 (Linearization of optimization problems)** Convert each of the following problems into a linear optimization problem.

(a) Let $\mathbf{x} \in \mathbb{R}^N$, $\mathbf{b} \in \mathbb{R}^M$, $\mathbf{A} \in \mathbb{R}^M \times \mathbb{R}^N$, and $\mathbf{c}_k = (c_{k1}, ..., c_{kN})$, $k \in \{1, ..., K\}$. Assume that $\mathbf{A}$, $\mathbf{b}$, and $\mathbf{c}_k$s are given.

$$\text{minimize max } (\mathbf{c}_1^T \mathbf{x}, ..., \mathbf{c}_K^T \mathbf{x})$$

$$\text{subject to } \mathbf{Ax} \ge \mathbf{b}$$

**HINT:** Introduce an additional variable and enforce it to be at least $\mathbf{c}_k^T \mathbf{x}$ for all $k$.

(b) Let $\mathbf{x} \in \mathbb{R}^N$, $\mathbf{b} \in \mathbb{R}^M$, $\mathbf{A} \in \mathbb{R}^M \times \mathbb{R}^N$, and $\mathbf{c} \in \mathbb{R}^N$. Assume that $\mathbf{A}$, $\mathbf{b}$, and $\mathbf{c}$ are given.

$$\text{minimize } |\mathbf{c}^T \mathbf{x}|$$

$$\text{subject to } \mathbf{Ax} \ge \mathbf{b}$$

(c) Let $\mathbf{x} \in \mathbb{R}^N$, $\mathbf{b} \in \mathbb{R}^M$, $\mathbf{A} \in \mathbb{R}^M \times \mathbb{R}^N$, and $\mathbf{c} \in \mathbb{R}^N$. Assume that $\mathbf{A}$, $\mathbf{b}$, and $\mathbf{c}$ are given.

$$\text{minimize} \sum_{i=1}^{N} c_i |x_i|$$

subject to $\mathbf{Ax} \geq \mathbf{b}$

**Problem 3.4 (Simplex algorithm)** Consider the following linear optimization problem.

$$\text{minimize} - x_1 - 2x_2$$
$$\text{subject to} - x_1 + x_2 \leq 2$$
$$x_1 + x_2 \leq 4$$
$$x_1, x_2 \geq 0$$

(a) Convert the problem to the standard form and construct the BFS in which $x_1$ and $x_2$ are nonbasic variables.

(b) Carry out the full tableau implementation of the simplex algorithm, starting with the BFS of part (a). Specify the optimal solution and the associated optimal cost.

(c) Draw the feasible set of the problem in terms of the original variables $x_1$ and $x_2$, i.e. two-dimensional domain set. Indicate the path taken by the simplex algorithm.

**Problem 3.5 (Two-phase simplex algorithm)** Consider the following linear optimization problem.

$$\text{minimize} \ 3x_1 + 4x_2 + 5x_3$$
$$\text{subject to} \ x_1 + x_2 + x_3 = -1$$
$$- x_1 - x_2 = 5$$
$$x_1, x_2, x_3 \geq 0$$

(a) Perform the two-phase simplex algorithm to solve the given problem. Conclude if the problem is feasible, unbounded, or has an optimal solution.

(b) Is there a basic solution for the given problem?

**Problem 3.6 (Duality of linear optimization)** Consider a linear optimization problem in the general form, i.e.

$$\text{minimize } \mathbf{c}^T\mathbf{x}$$

$$\text{subject to } \mathbf{Ax} \geq \mathbf{b}$$

(a) Viewing the problem as a convex optimization problem, write down the Lagrangian for the problem.

(b) Write down the dual function in terms of $\mathbf{c}$, $\mathbf{A}$, $\mathbf{b}$, and the dual variables.

(c) Write down the dual problem in terms of $\mathbf{c}$, $\mathbf{A}$, $\mathbf{b}$, and the dual variables.

(d) Use the result from part (c) to write down the dual problem of the following primal problem.

$$\text{minimize } x_1 + 2x_2$$

$$\text{subject to } x_1 + x_2 \geq 1$$

$$x_1 \geq -1$$

$$-x_1 + x_2 \geq -4$$

**Problem 3.7 (Duality of dual problem)** Consider the following linear optimization problem in the standard form.

$$\text{minimize } c_1x_1 + c_2x_2$$

$$\text{subject to } a_{11}x_1 + a_{12}x_2 = b_1$$

$$a_{21}x_1 + a_{22}x_2 = b_2$$

$$x_1, x_2 \geq 0$$

(a) Write down the corresponding dual problem.

(b) Express the dual problem in part (a) in the standard form.

(c) Write down the corresponding dual problem of the standard-form problem in part (b), i.e. find the dual of the dual problem.

(d) Show that the problem in part (c) can be expressed as the original primal problem.

**Problem 3.8 (Minimum cost routing with limited budget)**   Consider the minimum cost routing problem as given in (3.17). In particular, consider solving the problem for the network topology as shown in fig. 3.8. Note that there are 5 nodes and 14 directional links.

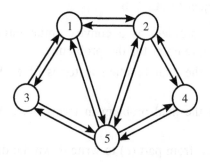

**Figure 3.8**   Network topology for problem 3.8.

Assume that $\alpha_l = 1$ and $c_l = 5$ for all links. In addition, assume that there are six s-d pairs with nonzero traffic, as shown below together with the demands and the sets of candidate paths.

| s-d pair | Demand | Candidate paths |
|----------|--------|-----------------|
| 1-3 | 1 | $1 \to 2 \to 3,\ 1 \to 4 \to 3$ |
| 2-4 | 2 | $2 \to 4,\ 2 \to 1 \to 4$ |
| 3-1 | 3 | $3 \to 2 \to 1,\ 3 \to 4 \to 1$ |
| 3-5 | 1 | $3 \to 4 \to 5,\ 3 \to 2 \to 1 \to 5$ |
| 4-2 | 2 | $4 \to 2,\ 4 \to 3 \to 2$ |
| 4-5 | 3 | $4 \to 5,\ 4 \to 1 \to 5$ |

(*a*) Write down the number of variables and the number of constraints used to solve the problem for the given scenario.

(*b*) Suppose that there is a limited budget $B > 0$ for the total cost of using network links. More specifically, given all the flow values $x^p$'s, the total cost is $\sum_{l \in \mathcal{L}} \alpha_l \left( \sum_{p \in P_l} x^p \right)$. With the limited budget, not all traffic demands may be supported. In addition, suppose that there is a revenue of $\beta^s$ gained for each unit of traffic supported for s-d pair $s$.

Reformulate the routing problem to maximize the total revenue gained given that the total cost of using network links cannot exceed $B$.

(c) Write down the number of variables and the number of constraints used to solve the problem in part (b) for the given scenario.

# Chapter 4

# Integer Linear Optimization

This chapter focuses on optimization problems in which the objective and constraint functions are linear while decision variables must take integer values. Such problems are called *integer linear optimization* problems. Compared to linear optimization, integer linear optimization problems are in general much more difficult to solve. In practice, heuristics are often needed to obtain approximate solutions. While duality theory also exists for integer linear optimization, strong duality does not always hold (unlike for linear optimization). After the discussion on the fundamentals of integer linear optimization, two example applications are discussed: routing and wavelength assignment in optical networks, and network topology design.

## 4.1 Illustrative Examples

The general form of an integer linear optimization problem looks the same as the linear optimization problem in (3.2), except for the additional integer constraint as shown below.[1]

---

[1] Let $\mathbb{Z}$ denote the set of all integers, and $\mathbb{Z}^+$ denote the set of all nonnegative integers.

$$\boxed{\begin{array}{c} \text{minimize } \mathbf{c}^{\mathsf{T}}\mathbf{x} \\ \text{subject to } \mathbf{Ax} \geq \mathbf{b} \\ \mathbf{x} \in \mathbb{Z}^{N} \end{array}} \qquad (4.1)$$

To illustrate the difficulty associated with integer linear optimization, consider the examples shown in figure 4.1. Suppose that the integer constraint $\mathbf{x} \in \mathbb{Z}^{N}$ is ignored and the corresponding linear optimization problem is solved. Such a process is called *relaxation*, while the corresponding linear optimization problem is called the *relaxed* problem. There are two possibilities.

1. The optimal solution obtained from the relaxed problem is an integer point, as illustrated in figure 4.1(*a*). In this case, since no noninteger point provides a lower cost than this integer point, the optimal solution from the relaxed problem is also optimal for the integer linear optimization problem.

2. The optimal solution obtained from the relaxed problem is not an integer point, as illustrated in figure 4.1(*b*). In this case, the optimal cost obtained from the relaxed problem (denoted by $f_{\text{relax}}$) yields a *lower bound* on the optimal cost of the integer linear optimization problem (denoted by $f^{*}$).

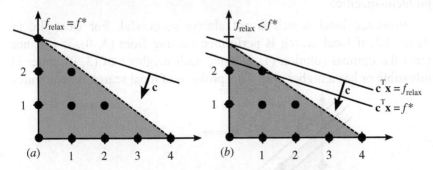

**Figure 4.1** Two examples of integer relaxation. In each figure, the shaded region indicates the feasible set of the relaxed problem.

It is interesting to note that, for some integer linear optimization problems, all the extreme points of the feasible set are integer points. For these problems, an optimal solution can be found by solving the corresponding relaxed problems. Such problems are called *unimodular* and are relatively easy to solve [Papadimitriou and Steiglitz, 1998]. Unfortunately, most integer linear optimization problems in practice do not have this unimodular property.

Before discussing a method to obtain exact optimal solutions for integer linear optimization problems, consider getting approximate solutions through *integer rounding*, which refers to rounding the solution obtained from relaxation to the nearest integer point. There are two difficulties with integer rounding, as illustrated in figure 4.2.

1. Rounding to the nearest integer point is not always feasible, e.g. point (4, 0) in figure 4.2.

2. Rounding to the nearest feasible integer point is not always optimal, e.g. point (3, 0) in figure 4.2. Note that, in figure 4.2, the optimal integer point is (1, 2).

In any case, the solution obtained from integer rounding (denoted by $f_{rounding}$) provides an *upper bound* on the optimal cost. In summary, $f_{relax} \leq f^* \leq f_{rounding}$. For practical purposes, with $f_{relax}$ and $f_{rounding}$ sufficiently close, the obtained solution from integer rounding may be used.

An improvement on integer rounding can be made by checking the costs of the *neighbors* of the current solution and moving to a neighbor with a lower cost. Such a procedure is called *local search*. As an example, the set of neighbors of a point $(x_1, x_2)$ can contain four points: $(x_1 \pm 1, x_2)$ and $(x_1, x_2 \pm 1)$. In general, the definition of a neighbor is problem-specific.

However, local search is not always successful. For example, in figure 4.2, if local search is performed starting from (3, 0), it will not yield the optimal solution (1, 2) since each neighbor of (3, 0) either is infeasible or has a higher cost. This problem of local search is sometimes

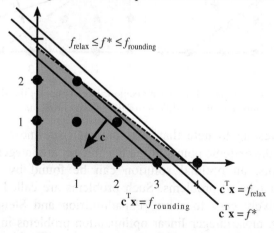

**Figure 4.2**  Illustration of relaxation followed by integer rounding.

described as getting stuck in a local but not global minimum. Later on in this chapter, a heuristic that deals with the local minimum problem will be discussed.

For several integer linear optimization problems, the feasible set contains a finite number of integer points. Because of this finiteness, one may consider performing *exhaustive search*, which refers to checking all possible solutions. The problem with exhaustive search is that the size of the feasible set usually grows exponentially with the problem size, making it computationally tractable only for small to moderate problem sizes.

Finally, while problem-specific algorithms have been developed to solve several integer linear optimization problems, there are still a large number of practical problems in which no one has found efficient algorithmic solutions. More specifically, these problems belong to the class of *NP-complete problems*, which remain open areas for research. NPcompleteness will not be discussed in this book; more information can be found in [Garey and Johnson, 1979].

In summary, the above discussions imply that, for integer linear optimization, algorithms that yield exact optimal solutions may have limited use in practice due to high computational complexity. Therefore, the development of heuristics for approximated solutions is an important research area for integer linear optimization.

## 4.2 Branch-and-Bound

*Branch-and-bound* is commonly used to obtain exact optimal solutions for integer linear optimization. It is based on the divide-and-conquer approach in problem solving. Roughly speaking, if a problem is difficult to solve, it is divided into several smaller sub-problems. Each sub-problem is then considered and may be divided further into smaller sub-problems if it is still difficult to solve.

Consider an integer linear optimization problem in the standard form shown below.

$$\text{minimize } \mathbf{c}^T\mathbf{x}$$

$$\text{subject to } \mathbf{Ax} = \mathbf{b}$$

$$\mathbf{x} \geq \mathbf{0}, \mathbf{x} \in \mathbb{Z}^N \tag{4.2}$$

Let $\mathcal{F}$ denote the feasible set of problem 4.2. If minimizing $\mathbf{c}^T\mathbf{x}$ subject to $\mathbf{x} \in \mathcal{F}$ is difficult to solve, then $\mathcal{F}$ can be partitioned into subsets $\mathcal{F}_1, ..., \mathcal{F}_K$. The sub-problem for each $\mathcal{F}_i$, i.e. minimizing $\mathbf{c}^T\mathbf{x}$ subject to $\mathbf{x} \in \mathcal{F}_i$, is then considered. If the sub-problem for $\mathcal{F}_i$ is still difficult, $\mathcal{F}_i$ can be further partitioned into subsets $\mathcal{F}_{i,1}, ..., \mathcal{F}_{i,K_i}$, and so on.

Compared to integer linear optimization, linear optimization is considered not difficult. Accordingly, problem 4.2 can be considered not difficult if its relaxation, i.e. minimizing $\mathbf{c}^T\mathbf{x}$ subject to $\mathbf{Ax} = \mathbf{b}$ and $\mathbf{x} \geq \mathbf{0}$, yields an integer optimal point.

In what follows, an unsolved sub-problem is referred to as an *active* sub-problem. An active sub-problem will be described by its feasible set. To describe the branch-and-bound procedure, the following notations are defined.

- $\mathcal{A}$: set of active sub-problems
- $\mathcal{F}_i$: feasible set of sub-problem $i$
- $\mathbf{x}_{\text{relax}}^i$: optimal solution obtained from relaxation with respect to sub-problem $i$
- $f_{\text{relax}}^i$: optimal cost obtained from relaxation with respect to sub-problem $i$
- $\mathbf{x}_{\text{best}}$: best integer solution found so far
- $f_{\text{best}}$: best cost found so far from an integer solution

An iteration of branch-and-bound is described next.

### Iteration of Branch-and-Bound

Initialize $\mathcal{A} = \{\mathcal{F}\}$ and $f_{\text{best}} = \infty$.

1. Remove one active sub-problem, say sub-problem $i$, from $\mathcal{A}$.
2. If sub-problem $i$ is infeasible, i.e. $\mathcal{F}_i = \varnothing$, discard the sub-problem and go to step 5. Otherwise, compute $\mathbf{x}_{\text{relax}}^i$ and $f_{\text{relax}}^i$ through relaxation and go to step 3.
3. If $f_{\text{relax}}^i \geq f_{\text{best}}$, discard the sub-problem and go to step 5. Otherwise, go to step 4.

4. If $\mathbf{x}_{\text{relax}}^i$ is an integer point, set $\mathbf{x}_{\text{best}} = \mathbf{x}_{\text{relax}}^i$ and go to step 5. Otherwise, partition $\mathcal{F}_i$ into smaller subsets, add these subsets to $\mathcal{A}$, and go to step 5.

5. If $\mathcal{A} \neq \varnothing$, go to step 1. Otherwise, terminate by returning $\mathbf{x}_{\text{best}}$ and $f_{\text{best}}$ as an optimal solution and the optimal cost respectively.

In general, there is some freedom in proceeding with branch-and-bound. The sub-problems in $\mathcal{A}$ can be considered in different orders, e.g. breadth-first search or depth-first search in a tree structure containing all sub-problems. In addition, there are several methods of partitioning $\mathcal{F}_i$ into smaller subsets.

One simple method of partioning $\mathcal{F}_i$ is to use a noninteger component of $\mathbf{x}_{\text{relax}}^i$. In particular, suppose the $j$th component $x_{\text{relax}, j}^i$ is not an integer. Two sub-problems can be created by adding each of the two constraints

$$x_j \leq \left\lfloor x_{\text{relax}, j}^i \right\rfloor \text{ or } x_j \geq \left\lceil x_{\text{relax}, j}^i \right\rceil.$$

The next example illustrates operations of branch-and-bound under this method.

**Example 4.1:** Consider the following problem.

$$\text{minimize } x_1 - 2x_2$$

$$\text{subject to} - 4x_1 + 6x_2 \leq 9$$

$$x_1 + x_2 \leq 4$$

$$x_1, x_2 \geq 0$$

$$x_1, x_2 \in \mathbb{Z}$$

Initially, set $f_{\text{best}} = \infty$. The first iteration solves the relaxed problem to get the relaxed optimal solution $\mathbf{x}_{\text{relax}, 2}^1 = (1.5, 2.5)$ with the relaxed optimal cost $f_{\text{relax}}^1 = -3.5$. From $x_{\text{relax}, 2}^1 = 2.5$, suppose that two sub-problems are created from additional constraints: sub-problem 1 with $x_2 \geq 3$ and sub-problem 2 with $x_2 \leq 2$, as illustrated in figure 4.3. Note that the set of active sub-problems is now $\mathcal{A} = \{\mathcal{F}_1, \mathcal{F}_2\}$, where $\mathcal{F}_1$ and $\mathcal{F}_2$ are the feasible sets of sub-problems 1 and 2 respectively.

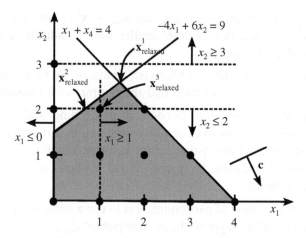

**Figure 4.3**    Iterations of branch-and-bound.

Since sub-problem 1 (with additional constraint $x_2 \geq 3$) is infeasible, it is deleted. By relaxing sub-problem 2, $\mathbf{x}^2_{\text{relax}} = (0.75, 2)$ and $f_{\text{relax}} = -3.25$. Sub-problem 2 is further partitioned into two further sub-problems: sub-problem 3 with $x_1 \geq 1$ and sub-problem 4 with $x_1 \leq 0$. Note that the set of active sub-problems is now $\mathcal{A} = \{\mathcal{F}_3, \mathcal{F}_4\}$, where $\mathcal{F}_3$ and $\mathcal{F}_4$ are the feasible sets of sub-problems 3 and 4 respectively.

By relaxing sub-problem 3, $\mathbf{x}^3_{\text{relax}} = (1, 2)$ and $f^3_{\text{relax}} = -3$. Since $\mathbf{x}^3_{\text{relax}}$ is an integer point, $f_{\text{best}}$ and $\mathbf{x}_{\text{best}}$ are updated to $-3$ and $(1, 2)$ respectively. By relaxing sub-problem 4, $\mathbf{x}^4_{\text{relax}} = (0, 1.5)$ and $f^4_{\text{relax}} = -3$. Since $f^4_{\text{relax}} \geq f_{\text{best}}$, sub-problem 4 is deleted.

Since $\mathcal{A}$ is now empty, branch-and-bound terminates by concluding that the optimal cost is $f^* = f_{\text{best}} = -3$ while the associated optimal solution is $\mathbf{x}^* = \mathbf{x}_{\text{best}} = (1, 2)$.    □

## †4.3    Cutting-Plane Algorithm

The *cutting-plane algorithm* is another method for obtaining an exact optimal solution to an integer linear optimization problem. Consider an integer linear optimization problem in the standard form as given below.

$$\text{minimize } \mathbf{c}^T\mathbf{x}$$

$$\text{subject to } \mathbf{Ax} = \mathbf{b}$$

$$\mathbf{x} \in \mathbb{Z}^+$$

The cutting-plane algorithm contains the following steps.

1. Obtain an optimal solution $\tilde{\mathbf{x}}$ from the relaxed problem.

2. If $\tilde{\mathbf{x}}$ is an integer point, then $\tilde{\mathbf{x}}$ is optimal for the original problem. The algorithm then terminates.

3. Otherwise, add an additional inequality constraint to the relaxed problem such that it is satisfied by all feasible integer solutions, but not by $\tilde{\mathbf{x}}$. Go back to step 1.

One systematic method to add a constraint in step 3 is the *Gomory cutting-plane algorithm*. In this algorithm, the additional constraint is found from the simplex tableau. In particular, the optimal tableau for the relaxed problem contains the constraints

$$\tilde{\mathbf{x}}_B + \mathbf{B}^{-1}\mathbf{A}_N\,\tilde{\mathbf{x}}_N = \mathbf{B}^{-1}\mathbf{b}, \tag{4.3}$$

where $\tilde{\mathbf{x}}_B$ and $\tilde{\mathbf{x}}_N$ are vectors of basic and nonbasic variables, and $\mathbf{A}_N$ is the matrix containing nonbasic columns. Let $y_{ij}$ be the $i$th component of $\mathbf{B}^{-1}\mathbf{A}_j$, and $y_{i0}$ be the $i$th component of $\mathbf{B}^{-1}\mathbf{b}$. Since $\tilde{\mathbf{x}}$ is not an integer point, there is at least one constraint from (4.3) with $y_{i0}$ being fractional, i.e.

$$\tilde{x}_{B(i)} + \sum_{j \notin \mathcal{I}} y_{ij}\,\tilde{x}_j = y_{i0}, \tag{4.4}$$

where $\tilde{x}_{B(i)}$ is the $i$th basic variable of $\tilde{\mathbf{x}}$ and $\mathcal{I}$ is the set of indices for basic variables. Since $\tilde{x}_j \geq 0$ for all $j$,

$$\tilde{x}_{B(i)} + \sum_{j \notin \mathcal{I}} \lfloor y_{ij} \rfloor \tilde{x}_j \leq \tilde{x}_{B(i)} + \sum_{j \notin \mathcal{I}} y_{ij}\,\tilde{x}_j = y_{i0}.$$

In addition, for any feasible integer point $\mathbf{x}$,

$$x_{B(i)} + \sum_{j \notin \mathcal{I}} \lfloor y_{ij} \rfloor x_j \leq \lfloor y_{i0} \rfloor \tag{4.5}$$

since the left hand side is always an integer. The above inequality is satisfied by all feasible integer points but not by $\tilde{\mathbf{x}}$ since $\tilde{x}_j = 0$ for all $j \notin \mathcal{I}$ and thus $\tilde{x}_i = y_{i0} > \lfloor y_{i0} \rfloor$. Subtracting (4.5) from (4.4) with integer point $\mathbf{x}$ yields the *Gomory cut*

$$\sum_{j \notin \mathcal{I}} f_{ij} x_j \geq f_{i0}, \qquad (4.6)$$

where $f_{ij}$ is the fractional part of $y_{ij}$, i.e. $f_{ij} = y_{ij} - \lfloor y_{ij} \rfloor$.

After adding the Gomory cut to the set of constraints, the simplex tableau can be updated to find another optimal solution to the relaxed problem. Since changing the constraint of the primal problem does not change the feasibility of the dual simplex algorithm, it follows that the dual simplex algorithm is appropriate for the update.

**Example 4.2:** Consider the following integer linear optimization problem, which is transformed into a standard-form problem.

minimize $-2x_1 - x_2$          minimize $-2x_1 - x_2$

subject to $x_1 + 4x_2 \leq 12$          subject to $x_1 + 4x_2 + x_3 = 12$

$\qquad\qquad x_1 \leq 2 \qquad\qquad \Rightarrow \qquad\qquad x_1 + x_4 = 2$

$\qquad\qquad x_1, x_2 \in \mathbb{Z}^+ \qquad\qquad\qquad x_1, x_2, x_3, x_4 \in \mathbb{Z}^+$

The simplex algorithm is performed for the relaxed problem as follows.

Iteration 1

|        | $x_1$ | $x_2$ | $x_3$ | $x_4$ |
|------|------|------|------|------|
| 0    | $-2$ | $-1$ | 0    | 0    |
| $x_3 =$ 12 | 1 | 4 | 1 | 0 |
| $x_4 =$ 2 | $1^*$ | 0 | 0 | 1 |

Iteration 2

|        | $x_1$ | $x_2$ | $x_3$ | $x_4$ |
|------|------|------|------|------|
| 4    | 0    | $-1$ | 0    | 2    |
| $x_3 =$ 10 | 0 | $4^*$ | 1 | $-1$ |
| $x_1 =$ 2 | 1 | 0 | 0 | 1 |

Iteration 3

|        | $x_1$ | $x_2$ | $x_3$ | $x_4$ |
|------|------|------|------|------|
| 6.5  | 0    | 0    | 0.25 | 1.75 |
| $x_2 =$ 2.5 | 0 | 1 | 0.25 | $-0.25$ |
| $x_1 =$ 2 | 1 | 0 | 0 | 1 |

From the optimal simplex tableau of the relaxed problem, the Gomory cut is $0.25x_3 + 0.75x_4 \geq 0.5$, or equivalently $x_3 + 3x_4 \geq 2$. From the two equality constraints, i.e. $x_3 = 12 - x_1 - 4x_2$ and $x_4 = 2 - x_1$, the Gomory cut can be expressed as $x_1 + x_2 \leq 4$. Figure 4.4 illustrates the Gomory cut for the relaxed problem.

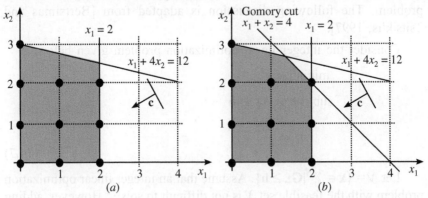

**Figure 4.4**   Operation of the Gomory cutting-plane algorithm.

Adding the constraint $x_3 + 3x_4 \geq 2$ or equivalently $-x_3 - 3x_4 + x_5 = -2$ with $x_5 \geq 0$, the dual simplex algorithm can be used to proceed further as follows. Note that the new variable $x_5$ can be taken as a basic variable.

Iteration 1

|        |      | $x_1$ | $x_2$ | $x_3$ | $x_4$ | $x_5$ |
|--------|------|-------|-------|-------|-------|-------|
|        | 6.5  | 0     | 0     | 0.25  | 1.75  | 0     |
| $x_2 =$ | 2.5  | 0     | 1     | 0.25  | −0.25 | 0     |
| $x_1 =$ | 2    | 1     | 0     | 0     | 1     | 0     |
| $x_5 =$ | −2   | 0     | 0     | $-1^*$ | − 3   | 1     |

Iteration 2

|        |      | $x_1$ | $x_2$ | $x_3$ | $x_4$ | $x_5$ |
|--------|------|-------|-------|-------|-------|-------|
|        | 6.5  | 0     | 0     | 0.25  | 1.75  | 0     |
| $x_3 =$ | 2    | 0     | 1     | 0     | −1    | 0.25  |
| $x_1 =$ | 2    | 1     | 0     | 0     | 1     | 0     |
| $x_5 =$ | 2    | 0     | 0     | 1     | 3     | −1    |

At this point, an optimal integer point $(x_1, x_2) = (2, 2)$ is obtained. It is interesting to note from figure 4.4 that $(2, 2)$ is not an extreme point of the original feasible set for relaxation. The Gomory cut $x_1 + x_2 \leq 4$ then makes $(2, 2)$ an extreme point for the relaxed problem.   □

## 4.4   Duality for Integer Linear Optimization

Unlike linear optimization, integer linear optimization does not have the strong duality property as stated in theorem 3.7. However, duality can be used to obtained a better lower bound compared to solving the relaxed problem. The following discussion is adapted from [Bertsimas and Tsitsiklis, 1997].

Consider the integer linear optimization problem given below.

$$\text{minimize } \mathbf{c}^T\mathbf{x}$$

$$\text{subject to } \mathbf{Ax} \geq \mathbf{b}$$

$$\mathbf{Gx} \geq \mathbf{h}$$

$$\mathbf{x} \in \mathbb{Z}^N \tag{4.7}$$

Let $\mathcal{Y} = \{\mathbf{x} \in \mathbb{Z}^N | \mathbf{Gx} \geq \mathbf{h}\}$. Assume that an integer linear optimization problem with the feasible set $\mathcal{Y}$ is not difficult to solve.[2] However, adding the constraint $\mathbf{Ax} \geq \mathbf{b}$ makes the problem difficult to solve. In this case, the Lagrangian can be written to contain the constraint $\mathbf{Ax} \geq \mathbf{b}$ as follows.

$$\Lambda(\mathbf{x}, \lambda) = \mathbf{c}^T\mathbf{x} + \lambda^T(\mathbf{b} - \mathbf{Ax})$$

The corresponding dual function $q(\lambda)$ is the optimal cost of the following problem.

$$\text{minimize } \mathbf{c}^T\mathbf{x} + \lambda^T(\mathbf{b} - \mathbf{Ax})$$

$$\text{subject to } \mathbf{x} \in \mathcal{Y} \tag{4.8}$$

It is tempting to use the weak duality result in chapter 3. However, since the constraint set $\mathcal{Y}$ is not convex, weak duality is reestablished below.

**Theorem 4.1 (Weak duality for integer linear optimization)**

Suppose that the integer linear optimization problem in (4.7) has a primal optimal solution with the optimal cost $f^*$.

1. For any $\lambda \geq \mathbf{0}$, $q(\lambda) \leq f^*$.
2. Let $q^* = \sup_{\lambda \geq 0} q(\lambda)$. Then, $q^* \leq f^*$.

---

[2]In general, some integer linear optimization problems are easy to solve using problem-specific (not general) algorithms, e.g. the well known Dijkstra algorithm for the shortest path routing problem. These algorithms can be used as subroutines for solving more difficult problems. This book does not discuss these special algorithms.

**Proof**

1. Let $\mathbf{x}^*$ denote a primal optimal solution. Since $\mathbf{b} - \mathbf{Ax}^* \leq \mathbf{0}$, for $\lambda \geq \mathbf{0}$,

$$\mathbf{c}^T\mathbf{x}^* + \lambda^T(\mathbf{b} - \mathbf{Ax}^*) \leq \mathbf{c}^T\mathbf{x}^* = f^*.$$

Since $\mathbf{x}^* \in \mathcal{Y}$, $q(\lambda) \leq \mathbf{c}^T\mathbf{x}^* + \lambda^T(\mathbf{b} - \mathbf{Ax}^*)$, yielding $q(\lambda) \leq f^*$.

2. Since $q(\lambda) \leq f^*$ for all $\lambda \geq \mathbf{0}$, $q^* = \sup_{\lambda \geq \mathbf{0}} q(\lambda) \leq f^*$. □

## Lagrangian Relaxation

The process of obtaining a lower bound from $q^*$ is called *Lagrangian relaxation*. For several problems, Lagrangian relaxation yields better lower bounds compared to relaxation by ignoring the integer constraints. In fact, if $\mathbf{A}$, $\mathbf{b}$, $\mathbf{c}$, $\mathbf{G}$, and $\mathbf{h}$ have integer entries, then it can be shown that $f_{\text{relax}} \leq q^* \leq f^*$, where $f_{\text{relax}}$ denotes the lower bound on $f^*$ obtained from the relaxed problem [Bertsimas and Tsitsiklis, 1997].

In general, it is possible that $q^* < f^*$, i.e. there is a duality gap. The following example illustrates a problem with duality gap.

**Example 4.3 (from [Bertsimas and Tsitsiklis, 1997])** Consider the following problem.

$$\text{minimize } 3x_1 - x_2$$

$$\text{subject to } x_1 - x_2 \geq -1$$

$$-x_1 + 2x_2 \leq 5$$

$$3x_1 + 2x_2 \geq 3$$

$$6x_1 + x_2 \leq 15$$

$$x_1, x_2 \in \mathbb{Z}^+$$

Figure 4.5(a) illustrates the feasible set of integer solutions. By inspection, the optimal integer point is $\mathbf{x}^* = (1, 2)$ with the optimal cost $f^* = 1$. Consider the relaxed problem whose feasible set is shown by the shaded area in figure 4.5(a). By inspection, the optimal solution of the relaxed problem is $\mathbf{x}_{\text{relax}} = (1/5, 6/5)$, with the optimal cost $f_{\text{relax}} = -3/5$.

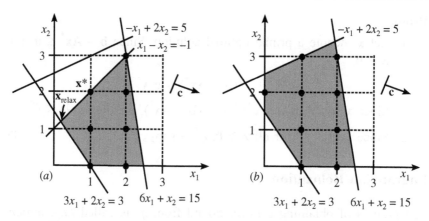

**Figure 4.5** Feasible sets for relaxation and Lagrangian relaxation.

Suppose that the first constraint $x_1 - x_2 \geq -1$ is relaxed. It follows that $\mathcal{Y} = \{(1, 0), (2, 0), (1, 1), (2, 1), (0, 2), (1, 2), (2, 2), (1, 3), (2, 3)\}$, as shown in figure 4.5(b). The dual function $q(\lambda)$ is given by $q(\lambda) = \min_{x \in \mathcal{Y}} 3x_1 - x_2 + \lambda(-1 - x_1 + x_2)$. Since there are nine integer points in $\mathcal{Y}$, $q(\lambda)$ is the minimum of nine linear functions, as illustrated in figure 4.6.

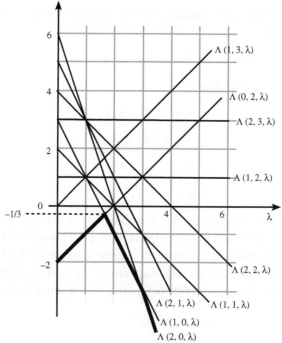

**Figure 4.6** Dual function $q(\lambda)$ for Lagrangian relaxation.

From figure 4.6, it follows that

$$q(\lambda) = \begin{cases} -2 + \lambda, & 0 \le \lambda \le 5/3 \\ 3 - 2\lambda, & 5/3 \le \lambda \le 3 \\ 6 - 3\lambda, & \lambda \ge 3 \end{cases}$$

yielding $q^* = -1/3$ at $\lambda^* = 5/3$. Note that $f_{\text{relax}} < q^* < f^*$ in this example. □

Lagrangian relaxation is also applicable in the presence of equality constraints. The only main difference is that dual variables for equality constraints can be negative. Since Lagrangian relaxation can yield a better lower bound than relaxation, it is often used in combination with branch and bound to obtain several efficient integer linear optimization algorithms [Bertsimas and Tsitsiklis, 1997].

# 4.5 Heuristics for Integer Linear Optimization

This section discusses two common methods for obtainting approximated optimal solutions for integer linear optimization problems. Methods for obtaining approximated solutions are referred to as *heuristics*. Several other heuristics exists, e.g. tabu search, genetic algorithm, particle swarm optimization; more information on these heuristics can be found in [Kennedy and Eberhart, 1995; Robertazzi, 1999].

## Local Search

*Local search* is a generic approach to obtain a *local optimum* for a general optimization problem. Consider the problem

$$\text{minimize } f(\mathbf{x})$$

$$\text{subject to } \mathbf{x} \in \mathcal{F}$$

where $\mathcal{F}$ denotes the feasible set. In general, local search can be applied even when the objective function $f$ is not linear and $\mathcal{F}$ is not an integer set. Starting with a point $\mathbf{x} \in \mathcal{F}$, local search proceeds by iterating as described below.

### *Iteration of Local Search*

1. Compute the cost $f(\mathbf{x})$ of the current solution $\mathbf{x}$.
2. Find a *neighbor* solution of $\mathbf{x}$, denoted by $\mathbf{y}$, such that $f(\mathbf{y}) < f(\mathbf{x})$. If such a $\mathbf{y}$ can be found, set $\mathbf{x} = \mathbf{y}$ and go back to step 1. Otherwise, local search terminates by returning the current $\mathbf{x}$ as the final solution.

To apply local search, we need to define what it means for one solution to be a neighbor of another solution. As a simple example, if $\mathcal{F}$ is a set of integer points, a neighbor of $(x_1, ..., x_N) \in \mathcal{F}$ can be one of the following points that is in $\mathcal{F}$: $(x_1 \pm 1, x_2, ..., x_N)$, $(x_1, x_2 \pm 1, ..., x_N)$, ..., $(x_1, x_2, ..., x_N \pm 1)$.

Based on the above neighbor definition, consider now a combination of integer rounding and local search for the example in figure 4.1($b$). Integer rounding of the optimal solution from relaxation yields (0, 2) as the starting feasible integer point for local search. Local search yields (1, 2) after one iteration. Since no neighbor of (1, 2) has a lower cost, local search terminates. In this case, local search found the optimal solution.

Consider again a combination of integer rounding and local search but for the example in figure 4.2. Integer rounding of the optimal solution from relaxation yields (3, 0) as the starting feasible integer point for local search. Local search yields (2, 0) after one iteration. Since no neighbor of (2, 0) has a lower cost, local search terminates. In this case, local search yields only a local minimum since (1, 2) is the global minimum.

In fact, the simplex algorithm for linear optimization can be considered as one kind of local search. Recall that the simplex algorithm starts with a basic feasible solution (BFS). A neighbor of a BFS is another BFS whose basic columns differ from the current BFS by only one column. In addition, moving from one BFS to its neighbor BFS can be done by the pivoting procedure. The simplex algorithm terminates after finding a local minimum, which is the same as the global minimum for linear optimization.

### [†]Simulated Annealing

*Simulated annealing* is a generic approach that tries to overcome the problem of getting stuck in a local but not global minimum in local search. It does so by allowing an occasional move to a neighbor with a strictly higher cost. The justification of simulated annealing can be done in the context of a Markov chain [Bertsimas and Tsitsiklis, 1997]. In particular, suppose that $\mathcal{F}$ is a finite set of feasible solutions. View each $\mathbf{x} \in \mathcal{F}$ as a state of a finite-state Markov chain. Denote the set of neighbors of state $\mathbf{x}$ by $\mathcal{N}(\mathbf{x})$. Accordingly, the set of feasible neighbors of $\mathbf{x}$ is $\mathcal{N}(\mathbf{x}) \cap \mathcal{F}$.

At the current state $\mathbf{x}$, a random feasible neighbor $\mathbf{y} \in \mathcal{N}(\mathbf{x}) \cap \mathcal{F}$ is selected with probability $q_{\mathbf{xy}}$. If $c(\mathbf{y}) \leq c(\mathbf{x})$, $\mathbf{y}$ becomes the current state. Otherwise, i.e. $c(\mathbf{y}) > c(\mathbf{x})$, $\mathbf{y}$ becomes the current state with probability

$$e^{-(c(\mathbf{y}) - c(\mathbf{x}))/T}, \tag{4.9}$$

where $T$ is a positive constant referred to as the *temperature*. The higher the temperature, the more likely the state transition from $\mathbf{x}$ to $\mathbf{y}$ will occur.

Let $\mathbf{x}^i$ be the state after $i$ transitions. Note that $\mathbf{x}^i$'s form a Markov chain with the set of states $\mathcal{F}$ and the probability of transition from state $\mathbf{x}$ to state $\mathbf{y} \in \mathcal{N}(\mathbf{x}) \cap \mathcal{F}$ given by the following expression.

$$p_{\mathbf{xy}} = \Pr\{\mathbf{x}^{i+1} = \mathbf{y}|\mathbf{x}^i = \mathbf{x}\} = \begin{cases} q_{\mathbf{xy}} e^{-(c(\mathbf{y}) - c(\mathbf{x}))/T}, & c(\mathbf{y}) > c(\mathbf{x}) \\ q_{\mathbf{xy}}, & \text{otherwise} \end{cases} \tag{4.10}$$

Assume that the above Markov chain is irreducible, which means that it is possible to move from any state $\mathbf{x}$ to any state $\mathbf{y}$ after a finite number of transitions. Define

$$\pi(\mathbf{x}) = e^{-c(\mathbf{x})/T}/A, \quad \text{where } A = \sum_{\mathbf{x} \in \mathcal{F}} e^{-c(\mathbf{x})/T}. \tag{4.11}$$

The role of $\pi(\mathbf{x})$ is stated in the following theorem. The justification is omitted but can be found in [Bertsimas and Tsitsiklis, 1997].

**Theorem 4.2**   Suppose that the Markov chain $\mathbf{x}^i$ is irreducible and $q_{\mathbf{xy}} = q_{\mathbf{yx}}$ for every pair of feasible neighbors $\mathbf{x}$ and $\mathbf{y}$. Then, the set $\{\pi(\mathbf{x})|\mathbf{x} \in \mathcal{F}\}$ is the unique set of steady-state probabilities of the Markov chain.

Based on the exponential decay of $\pi(\mathbf{x})$ as a function of cost $c(\mathbf{x})$, the following observations can be made. For a very small $T > 0$, the probability $\pi(\mathbf{x})$ is approximately zero except for the optimal states with the minimum value of $c(\mathbf{x})$. This means that simulated annealing will eventually approach an optimal state with high probability when $T$ is small. However, as a small $T$ makes it harder to escape from a local minimum, the algorithm may take quite a long time to reach an optimal state.

One common strategy is to use the *temperature schedule* of the form

$$T(i) = C/\log i, \; i > 1, \tag{4.12}$$

where $C$ is a large positive constant. Initially, a large $T$ will allow the algorithm to escape from a local minimum around the starting point. As the algorithm proceeds, a small $T$ will make it converge to an optimal solution with high probability.

Empirically, simulated annealing has been found to work reasonably well in several problems [Bertsimas and Tsitsiklis, 1997]. Being a generic approach, it cannot outperform specialized algorithms that are problem specific. Hence, it is appropriate to apply simulated annealing when the problem does not have any evident structure or property that can be exploited.

## 4.6  Application: RWA in a WDM Network

For networks utilizing optical transmissions, *wavelength division multiplexing (WDM)* refers to dividing the bandwidth available in an optical fiber into separate frequency channels normally referred to as *wavelength channels*. In most current commercial systems, WDM technologies are normally used for *point-to-point* transmissions. Switching of traffic is still performed using electronics, as illustrated in figure 4.7.

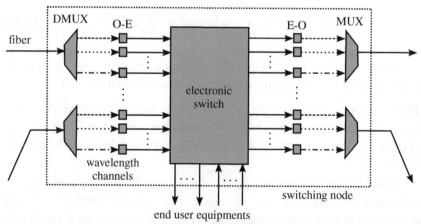

**Figure 4.7**  Electronic switching node architecture. E-O and O-E are electrical-to-optical converter (e.g. laser) and optical-to-electrical converter (e.g. photodiode) respectively. MUX and DMUX are optical wavelength multiplexer and demultiplexer respectively.

As optical switching technologies become more mature, the use of optical switching is expected in future WDM networks, as illustrated in figure 4.8. With optical switching, some traffic can bypass electronic switching at transit nodes that are neither the sources nor the destinations. This ability of bypassing electronic switching is referred to as *optical bypass* and is expected to provide significant cost savings from using a smaller amount of electronic equipments [Simmons, 2008].

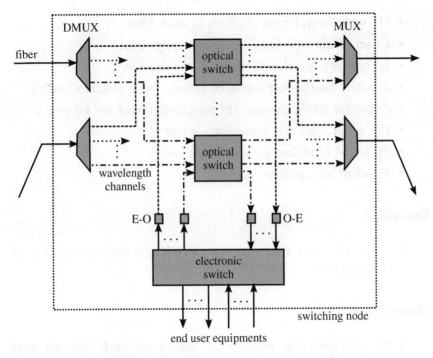

**Figure 4.8**   Hybrid optical-and-electronic switching node architecture. In general, more than one wavelength channel can be added/dropped to/from each optical switch.

However, a traffic stream that bypasses electronic switching needs to enter and leave the switching node on the same wavelength channel. This constraint is referred to as the *wavelength continuity constraint*, and gives rise to a lot of networking problems that are unique for WDM networks.

This section considers the static routing problem subject to the wavelength continuity constraint. This problem is known as the static *routing and wavelength assignment (RWA) problem*. The RWA problem is formulated as an integer linear optimization problem as follows. For convenience, the transmission rate of one wavelength channel is referred to as a wavelength unit.

## Given information

- $\mathcal{W}$: set of wavelength channels in each fiber
- $\mathcal{L}$: set of directed links (or equivalently fibers)
- $\alpha_l$: cost per wavelength channel in using link $l$
- $\mathcal{S}$: set of source-destination (s-d) pairs (with nonzero traffic)
- $t^s$: integer traffic demand (in wavelength unit) for s-d pair $s$
- $\mathcal{P}^s$: set of candidate paths for s-d pair $s$
- $\mathcal{P}_l$: set of candidate paths that use link $l$
- $\mathcal{P}$: set of all candidate paths

## Variables

- $x_w^p \in \{0, 1\}$: traffic flow (in wavelength unit) on path $p$ on wavelength $w$

## Constraints

- No collision (i.e. conflict of usage) on each link on each wavelength channel

$$\forall l \in \mathcal{L}, \forall w \in \mathcal{W}, \sum_{p \in \mathcal{P}_l} x_w^p \leq 1$$

- Satisfaction of traffic demands

$$\forall s \in \mathcal{S}, \sum_{p \in \mathcal{P}^s} \sum_{w \in \mathcal{W}} x_w^p = t^s$$

- Integer constraints

$$\forall p \in \mathcal{P}, \forall w \in \mathcal{W}, x_w^p \in \{0, 1\}$$

## *Objective*

- Minimize the total cost of link capacity usage

$$\text{minimize} \sum_{l \in L} \alpha_l \left( \sum_{p \in \mathcal{P}_l} \sum_{w \in \mathcal{W}} x_w^p \right)$$

The overall optimization problem is as follows.

$$\text{minimize} \sum_{l \in L} \alpha_l \left( \sum_{p \in \mathcal{P}_l} \sum_{w \in \mathcal{W}} x_w^p \right)$$

$$\text{subject to } \forall l \in \mathcal{L}, \forall w \in \mathcal{W}, \sum_{p \in \mathcal{P}_l} x_w^p \leq 1 \tag{4.13}$$

$$\forall s \in \mathcal{S}, \sum_{p \in \mathcal{P}^s} \sum_{w \in \mathcal{W}} x_w^p = t^s$$

$$\forall p \in \mathcal{P}, \forall w \in \mathcal{W}, x_w^p \in \{0, 1\}$$

**Example 4.4:** Consider the RWA problem for the network and traffic as shown in figure 4.9. Assume that $W = 2$ and $\alpha_l = 1$ for all $l$. The set of candidate paths are as follows.

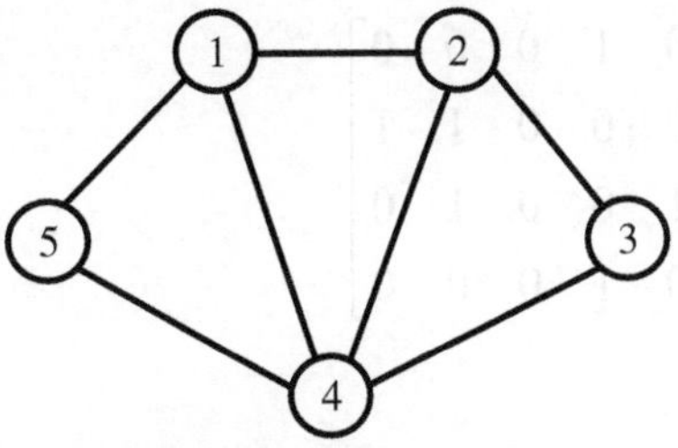

Each link is bidirectional.

$$[t^s] = \begin{array}{c} \\ \\ \text{Source of } s \end{array} \begin{array}{c} \\ 1 \\ 2 \\ 3 \\ 4 \\ 5 \end{array} \overset{\text{destination of } s}{\begin{array}{ccccc} 1 & 2 & 3 & 4 & 5 \\ \left[\begin{array}{ccccc} 0 & 0 & 1 & 0 & 0 \\ 0 & 0 & 0 & 3 & 0 \\ 1 & 0 & 0 & 0 & 1 \\ 0 & 2 & 0 & 0 & 1 \\ 0 & 0 & 1 & 2 & 0 \end{array}\right] \end{array}}$$

**Figure 4.9** Network topology and traffic RWA.

$$\mathcal{P}^{1-3} = \{1 \to 2 \to 3, 1 \to 4 \to 3\} \quad \mathcal{P}^{4-2} = \{4 \to 2, 4 \to 3 \to 2\}$$

$$\mathcal{P}^{2-4} = \{2 \to 4, 2 \to 1 \to 4\} \qquad \mathcal{P}^{4-5} = \{4 \to 5, 4 \to 1 \to 5\}$$

$$\mathcal{P}^{3-1} = \{3 \to 2 \to 1, 3 \to \qquad \mathcal{P}^{5-3} = \{5 \to 4 \to 3, 5 \to$$
$$4 \to 1\} \qquad\qquad\qquad 1 \to 2 \to 3\}$$

$$\mathcal{P}^{3-5} = \{3 \to 4 \to 5, 3 \to 2 \to \qquad \mathcal{P}^{5-4} = \{5 \to 4, 5 \to 1 \to 4\}$$
$$1 \to 5\}$$

The `glpk` command in Octave can be used to solve the RWA problem.[3] The obtained optimal RWA is shown below and is illustrated in figure 4.10. The corresponding optimal cost is 18.

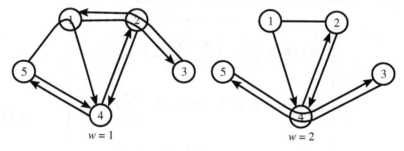

$w = 1$ $\qquad\qquad\qquad\qquad\qquad$ $w = 2$

**Figure 4.10** Illustrations of optimal RWAs.

$$\begin{bmatrix} x_1^{123*} & x_1^{143*} & x_1^{24*} & x_1^{214*} & x_1^{321*} & x_1^{341*} & x_1^{345*} & x_1^{3215*} \\ x_1^{42*} & x_1^{432*} & x_1^{45*} & x_1^{415*} & x_1^{543*} & x_1^{5123*} & x_1^{54*} & x_1^{514*} \\ x_2^{123*} & x_2^{143*} & x_2^{24*} & x_2^{214*} & x_2^{321*} & x_2^{341*} & x_2^{345*} & x_2^{3215*} \\ x_2^{42*} & x_2^{432*} & x_2^{45*} & x_2^{415*} & x_2^{543*} & x_2^{5123*} & x_2^{54*} & x_2^{514*} \end{bmatrix}$$

$$= \begin{bmatrix} 1 & 0 & 1 & 0 & 1 & 0 & 0 & 0 \\ 1 & 0 & 1 & 0 & 0 & 0 & 1 & 1 \\ 0 & 0 & 1 & 1 & 0 & 0 & 1 & 0 \\ 1 & 0 & 0 & 0 & 1 & 0 & 0 & 0 \end{bmatrix}$$

[3]See section C.3 in the appendix for specific Octave commands.

# 4.7 Application: Network Topology Design

Consider a *topology design* problem in which network node locations are given. The objective is to connect these nodes to form a connected network topology. Typically, the overall network topology is hierarchical, e.g. consisting of wide area networks (WANs), metro area networks (MANs), and local area networks (LANs).

Suppose that a two-layered architecture is considered. The first layer is a *backbone network* consisting of *concentrator nodes* selected from the given set of nodes. The second layer consists of multiple subnetworks each of which contains one concentrator node connected to non-concentrator nodes. Each non-concentrator node is connected to exactly one concentrator node. Figure 4.11 illustrates the overall two-layer network architecture.

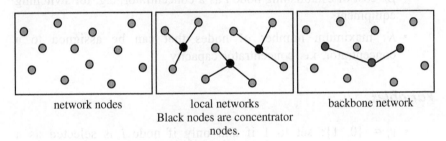

network nodes      local networks      backbone network
Black nodes are concentrator nodes.

**Figure 4.11** Two-layer network architecture.

More generally, the backbone network may have the survivability requirement. For example, the backbone topology should be such that any single link failure does not make the topology disconnected. For simplicity, survivability is not considered in what follows.

Consider breaking the problem of network topology design into the following two steps.[4]

1. Select concentrator nodes and assign each non-concentrator node to exactly one concentrator node. This problem will be referred to as the *concentrator location problem*.

---

[4]In general, breaking any problem into several steps or sub-problems may incur some loss of optimality. However, for several practical problems, breaking a problem into sub-problems makes the problem solvable within the available time; it also facilitates step-by-step development of relevant heuristics.

2. Connect the concentrator nodes to form a backbone network. This problem will be referred to as the *concentrator connection problem*.

## Concentrator Location Problem

The concentrator location problem can be formulated as an integer linear optimization problem as follows.

### Given information

- $N$: number of nodes
- $\alpha_{ij}$: cost of connecting node $i$ to node $j$
- $\beta_i$: cost of establishing node $i$ as a concentrator, e.g. for switching equipment.
- $K$: maximum number of nodes that can be assigned to a concentrator, i.e. concentrator capacity

### Variables

- $y_i \in \{0, 1\}$: set to 1 if and only if node $i$ is selected as a concentrator
- $x_{ij} \in \{0, 1\}$: set to 1 if and only if node $i$ is assigned to node $j$, which is assigned as a concentrator (assuming that $x_{ii} = 1$ if node $i$ is selected as a concentrator)

### Constraints

- Each node is connected to exactly one concentrator.

$$\forall i \in \{1, ..., N\}, \sum_{j=1}^{N} x_{ij} = 1$$

- Concentrator capacity

$$\forall j \in \{1, ..., N\}, \sum_{i=1}^{N} x_{ij} \leq K y_j$$

- Integer constraints

$$\forall i, j \in \{1, ..., N\}, x_{ij} \in \{0, 1\}$$
$$\forall i \in \{1, ..., N\}, y_i \in \{0, 1\}$$

## *Objective*

- Minimizing the total cost of establishing the network topology

$$\text{minimize} \sum_{i=1}^{N} \sum_{j=1}^{N} \alpha_{ij} x_{ij} + \sum_{i=1}^{N} \beta_i y_i$$

The overall integer linear optimization problem is shown below.

$$
\begin{array}{ll}
\text{minimize} & \displaystyle\sum_{i=1}^{N} \sum_{j=1}^{N} \alpha_{ij} x_{ij} + \sum_{i=1}^{N} \beta_i y_i \\[2ex]
\text{subject to} & \forall i \in \{1, ..., N\}, \displaystyle\sum_{j=1}^{N} x_{ij} = 1 \\[2ex]
& \forall j \in \{1, ..., N\}, \displaystyle\sum_{j=1}^{N} x_{ij} \leq K y_j \\[2ex]
& \forall i, j \in \{1, ..., N\}, x_{ij} \in \{0, 1\} \\[1ex]
& \forall i \in \{1, ..., N\}, y_i \in \{0, 1\}
\end{array}
\tag{4.14}
$$

**Example 4.5:** Consider the concentrator location problem for the network shown in figure 4.12. Assume that $\alpha_{ij}$ is equal to the distance between nodes $i$ and $j$, $\beta_i = 10$ for all $i$, and $K = 3$.

The `glpk` command in Octave can be used to solve for an optimal solution.[5] The obtained optimal solution is shown below; the corresponding optimal cost is 26.7.

---

[5]See section C.4 in the appendix for specific Octave commands.

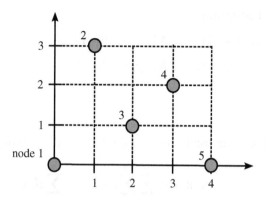

**Figure 4.12**    Node locations for the concentrator location problem.

$$
\begin{bmatrix}
x_{11}^* & x_{12}^* & x_{13}^* & x_{14}^* & x_{15}^* \\
x_{21}^* & x_{22}^* & x_{23}^* & x_{24}^* & x_{25}^* \\
x_{31}^* & x_{32}^* & x_{33}^* & x_{34}^* & x_{35}^* \\
x_{41}^* & x_{42}^* & x_{43}^* & x_{44}^* & x_{45}^* \\
x_{51}^* & x_{52}^* & x_{53}^* & x_{54}^* & x_{55}^* \\
y_1^* & y_2^* & y_3^* & y_4^* & y_5^*
\end{bmatrix}
=
\begin{bmatrix}
1 & 0 & 0 & 0 & 0 \\
0 & 0 & 0 & 1 & 0 \\
1 & 0 & 0 & 0 & 0 \\
0 & 0 & 0 & 1 & 0 \\
0 & 0 & 0 & 1 & 0 \\
1 & 0 & 0 & 1 & 0
\end{bmatrix}
$$

Suppose now that the concentrator establishment cost is $\beta_i = 3$ for all $i$. The new optimal solution is shown below; the corresponding optimal cost is 12.7.

$$
\begin{bmatrix}
x_{11}^* & x_{12}^* & x_{13}^* & x_{14}^* & x_{15}^* \\
x_{21}^* & x_{22}^* & x_{23}^* & x_{24}^* & x_{25}^* \\
x_{31}^* & x_{32}^* & x_{33}^* & x_{34}^* & x_{35}^* \\
x_{41}^* & x_{42}^* & x_{43}^* & x_{44}^* & x_{45}^* \\
x_{51}^* & x_{52}^* & x_{53}^* & x_{54}^* & x_{55}^* \\
y_1^* & y_2^* & y_3^* & y_4^* & y_5^*
\end{bmatrix}
=
\begin{bmatrix}
1 & 0 & 0 & 0 & 0 \\
0 & 0 & 1 & 0 & 0 \\
0 & 0 & 1 & 0 & 0 \\
0 & 0 & 1 & 0 & 0 \\
0 & 0 & 0 & 0 & 1 \\
1 & 0 & 1 & 0 & 1
\end{bmatrix}
$$

Notice that, a smaller concentrator establishment cost results in a larger number of concentrators.                                                    □

# †Concentrator Connection Problem

The objective of the concentrator connection problem is to connect all concentrator nodes using the minimum cost. Let $\gamma_{ij}$ denote the cost of establishing a bidirectional link between nodes $i$ and $j$. The concentrator connection problem is then equivalent to finding a *minimum spanning tree (MST)*.

While the problem of finding an MST can be formulated as an optimization problem, there exists efficient algorithms for this purpose [Bertsekas and Gallager, 1992]. Thus, the problem of finding an MST provides an example in which problem specific algorithms can be developed to obtain an optimal solution.

The *Prim-Dijkstra algorithm* starts with an arbitrary node as a sub-tree and iteratively adds a new link with the minimum cost to enlarge the sub-tree until all nodes are contained in the sub-tree. The resultant sub-tree is a MST. Figure 4.13 illustrates the operations of the Prim-Dijkstra algorithm.

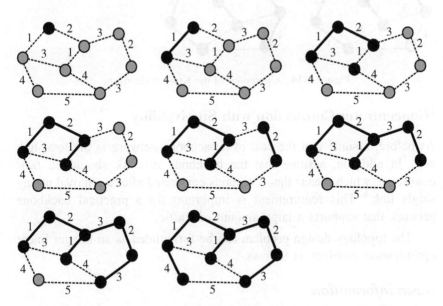

**Figure 4.13**  Operations of the Prim-Dijkstra algorithm. The first node of the sub-tree is colored black in the first diagram. Each link label indicates the cost of establishing that link.

The *Kruskal algorithm* starts with each node being a substree and iteratively combines two subtrees to form a bigger sub-tree using a link with the minimum cost; the iteration stops when there is a single sub-tree left. The resultant sub-tree is a MST. Figure 4.14 illustrates the operations of the Kruskal algorithm.

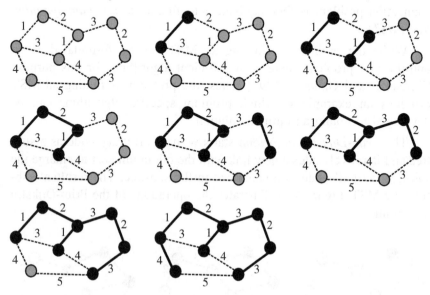

**Figure 4.14**   Operations of the Kruskal algorithm.

## †Concentrator Connection with Survivability

As before, assume that the cost of a backbone network is the total link cost. In addition, assume that the backbone network should be *two-connected*, which means that it remains connected after a removal of any single link.[6] This requirement is important for a practical backbone network that supports a large amount of traffic.

The topology design problem can be formulated as an integer linear optimization problem as follows.

### Given information

- $\mathcal{N}$: set of nodes
- $\mathcal{P}(\mathcal{N})$: set of all subsets (i.e. power set) of $\mathcal{N}$

---

[6]More generally, a network topology is *k-connected* if it remains connected after removing any $k-1$ links.

- $\gamma_{ij}$ : cost of creating a directed link from node $i$ to node $j$ (equal to 0 for $i = j$)

## Variables

- $x_{ij} \in \{0, 1\}$: equal to 1 if and only if node $i$ is connected to node $j$

## Constraints

- Two-connectedness constraint[7]

$$\forall S \in \mathcal{P}(\mathcal{N}) \text{ with } S \neq \emptyset, \mathcal{N}, \sum_{i \in S} \sum_{j \notin S} x_{ij} \geq 2$$

- Integer constraints

$$\forall i, j \in \mathcal{N}, x_{ij} \in \{0, 1\}$$

## Objective

- Minimize the total cost of a network topology

$$\text{minimize } \sum_{i \in N} \sum_{j \in N} \gamma_{ij} x_{ij}$$

The overall optimization problem is as follows.

$$\begin{array}{ll}
\text{minimize} & \displaystyle\sum_{i \in N} \sum_{j \in N} \gamma_{ij} x_{ij} \\[2em]
& \forall S \in \mathcal{P}(\mathcal{N}) \text{ with } S \neq \emptyset, \mathcal{N}, \displaystyle\sum_{i \in S} \sum_{j \in S} x_{ij} \geq 2 \qquad (4.15) \\[2em]
& \forall i, j \in \mathcal{N}, x_{ij} \in \{0, 1\}
\end{array}$$

---

[7]Note that a network topology is two-connected if and only if there are at least two links from each nonempty subset of nodes to its nonempty complement. If a $k$-connected network topology is required, the value of 2 in the constraint can be replaced by $k$.

While the problem description is rather compact, the associated computational complexity is high since the size of $\mathcal{P}(\mathcal{N})$ grows exponentially with the number of network nodes.

## 4.8  Exercise Problems

**Problem 4.1 (Problem formulation)** Consider a problem of establishing communication links between one source-destination pair. Suppose that the total data rate to be supported is $R$ (in bps). Assume that there are $M$ types of transmission lines. A type-$m$ line, where $m \in \{1, ..., M\}$, can support up to $g_m$ (in bps) and costs $c_m$ (in dollars per month).

Note that $R$, $M$, $g_m$'s, and $c_m$'s are given and are not decision variables. The problem objective is to determine the number of type-$m$ lines for each $m \in \{1, ..., M\}$ so that the total cost of transmission lines is minimized while the data rate of $R$ is supported.

Formulate the problem as an integer linear optimization problem. Assume that $R$, $M$, $g_m$'s, and $c_m$'s are nonnegative integers.

**Problem 4.2 (Problem formulation)** Consider a problem of scheduling final exams for a group of $N$ students and $M$ courses. The following information is given.

- $a_{ij} \in \{0, 1\}, i \in \{1, ..., N\}, j \in \{1, ..., M\}$: equal to 1 if and only if student $i$ takes course $j$

Suppose that each day can only hold three-hour exams during 9:00-12:00. The problem objective is to schedule the $M$ exams using the minimum number of days subject to the constraint that no student has more than one exam on the same day. Note that two or more exams can be held on the same day if no student takes more than one of these courses.

Formulate the problem as an integer linear optimization problem. HINT: Start by assuming that there are $M$ days available so that the problem is always feasible. In addition, the following decision variables can be used.

- $y_k \in \{0, 1\}, k \in \{1, ..., M\}$: equal to 1 if and only if there is at least one exam on day $k$

- $x_{jk} \in \{0, 1\}, j, k \in \{1, ..., M\}$: equal to 1 if and only if the exam for course $j$ is scheduled on day $k$

**Problem 4.3 (Problem formulation)** Consider a cellular mobile system in which the service area is divided into multiple cells, as illustrated in figure 4.15.

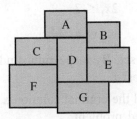

**Figure 4.15** Example cells in a service area for problem 4.3.

To support traffic in each cell, a set of carrier frequencies must be provided. However, due to signal interference, the same set of frequencies cannot be assigned to cells that are adjacent to each other. For example, in figure 4.15, cell A is adjacent to cell C. Therefore, cell A cannot use the same set of frequencies as for cell C. Similarly, cell A cannot use the same set of frequencies as for cell B. However, cell B and cell C are not adjacent and may use the same set of frequencies.

Suppose there are $N$ cells and $M$ set of frequencies. In addition, suppose that the following parameters are given.

- $a_{ij} \in \{0, 1\}, i, j \in \{1, ..., N\}$: equal to 1 if and only if cell $i$ and cell $j$ are adjacent

(a) Formulate an optimization problem to assign one set of frequencies to each cell. The objective is to minimize the number of frequency sets used. The constraints are (1) each cell must be assigned exactly one set of frequencies, and (2) no pair of adjacent cells are assigned the same set of frequencies.

(b) In terms of $N$ and $M$, specify the number of variables and the number of constraints (not including integer constraints) for the problem in part (a).

**Problem 4.4 (Branch-and-bound and Lagrangian relaxation)** Consider the following optimization problem.

$$\text{maximize } 2x_1 + x_2$$
$$\text{subject to } -2x_1 - 2x_2 \le -1$$
$$2x_1 + 2x_2 \le 5$$
$$-x_1 + 2x_2 \le 2$$
$$2x_1 \le 3$$
$$x_1, x_2 \in \mathbb{Z}^+$$

(a) Draw the feasible set. By inspection, find the optimal solution $\mathbf{x}^*$ and the optimal cost $f^*$.

(b) By inspection, find the optimal solution $\mathbf{x}_{\text{relax}}$ and the optimal cost $f_{\text{relax}}$ for the relaxed problem.

(c) Solve the problem by branch-and-bound. Each relevant relaxed problem can be solved by inspection.

(d) Suppose the constraint $2x_1 \le 3$ is relaxed. What is the dual optimal cost $q^*$ obtained from this Lagrangian relaxation?

**Problem 4.5 (RWA for maximizing revenue)**  Consider the routing and wavelength assignment (RWA) problem for a WDM network in which the main objective is to maximize the revenue. It is not required that all the traffic demands must be supported. The problem is formulated as follows.

## Given information

- $\mathcal{W}$: set of wavelength channels in each fiber
- $\mathcal{L}$: set of directed links (or equivalently fibers)
- $\mathcal{S}$: set of source-destination (s-d) pairs (with nonzero traffic)
- $t^s$: integer traffic demand (in wavelength unit) for s-d pair $s$
- $r^s > 0$: revenue gained from each supported traffic unit (in wavelength unit) for s-d pair $s$
- $\mathcal{P}^s$: set of candidate paths for s-d pair $s$
- $\mathcal{P}_l$: set of candidate paths that use link $l$
- $\mathcal{P}$: set of all candidate paths

## Variables

- $x_w^p \in \{0, 1\}$: traffic flow (in wavelength unit) on path $p$ on wavelength $w$

## Constraints

- No collision (i.e. conflict of usage) on each link on each wavelength channel

$$\forall l \in \mathcal{L}, \forall w \in \mathcal{W}, \sum_{p \in \mathcal{P}_l} x_w^p \leq 1$$

- Bounds on supported traffic due to traffic demands

$$\forall s \in \mathcal{S}, \sum_{p \in \mathcal{P}^s} \sum_{w \in \mathcal{W}_l} x_w^p \leq t^s$$

- Integer constraints

$$\forall p \in \mathcal{P}, \forall w \in \mathcal{W}, x_w^p \in \{0, 1\}$$

## Objective

- Maximize the total revenue

### Missing objective (to be specified)

(a) Write down the objective that is still missing.

(b) Count the number of variables and the number of constraints (not including the integer contraints of $x_w^p$) in terms of $|\mathcal{L}|$, $|\mathcal{W}|$, $|\mathcal{S}|$, and $|\mathcal{P}|$.

   NOTE: $|\mathcal{X}|$ denotes the number of elements in set $\mathcal{X}$. For example, if $\mathcal{X} = \{10, 20, 30\}$, then $|\mathcal{X}| = 3$.

(c) Does an optimal solution always exist? Why or why not?

(d) Let $\alpha_l > 0$ be the cost per wavelength channel in using link $l$. Suppose that the objective is to maximize the profit, which is equal to total revenue subtracted by the total cost of using wavelength channels. Write down the modified objective function.

(*e*) Let $\hat{\mathbf{x}}$ denote an optimal solution obtained from part (*a*), and $\tilde{\mathbf{x}}$ denote an optimal solution obtained from part (*d*). Consider a solution **x** whose components are given by

$$x_w^p = \min\,(\hat{x}_w^p, \tilde{x}_w^p).$$

Is **x** always feasible?

**Problem 4.6 (Solving a Sudoku problem)**   Consider solving a Sudoku problem using integer linear optimization. An example Sudoku problem is shown in figure 4.16. It consists of a $9 \times 9$ table with 9 rows and 9 columns. Each table entry is a digit in the set $\{1, ..., 9\}$. A valid solution of a Sudoku problem must satisfy the following constraints.

- Each digit $i \in \{1, ..., 9\}$ appears exactly once in each row.
- Each digit $i \in \{1, ..., 9\}$ appears exactly once in each column.
- Each digit $i \in \{1, ..., 9\}$ appears exactly once in each $3 \times 3$ sub-tables. Note that there are 9 such sub-tables, as shown in figure 4.16, with shaded and unshaded regions.

(*a*) Suppose that Alice has been formulating the above Sudoku problem as an integer linear optimization problem. While Alice's work has been correct, it is not yet finished. Specify the missing constraint expressions so that the problem formulation is complete.

## Variables

$x_{ijk} \in \{0, 1\}$: equal to 1 if and only if digit $i$ appears in row $j$ and in column $k$

## Given information

The given information consists of the already given entries in the Sudoku table. For example, in figure 4.16, $x_{411} = 1$, $x_{631} = 1$, $x_{942} = 1$, and so on.

## Objective

Since only a feasible solution is needed, any objective function will do. Hence, one possibility is to set a constant objective function as follows.

minimize 0

| 4 |   |   |   | 5 | 7 |   |   |   |
|---|---|---|---|---|---|---|---|---|
|   |   |   |   | 9 | 1 |   |   |   |
| 6 |   |   |   | 8 | 3 |   |   |   |
|   | 9 |   | 6 |   |   | 2 |   |   |
|   | 5 |   | 3 |   |   | 8 |   |   |
|   | 3 |   | 4 |   |   | 6 |   |   |
|   |   | 3 | 7 |   |   |   |   | 9 |
|   |   | 2 | 1 |   |   |   |   |   |
|   |   | 8 | 2 |   |   |   |   | 5 |

| 4 | 8 | 1 | 3 | 2 | 5 | 7 | 9 | 6 |
|---|---|---|---|---|---|---|---|---|
| 3 | 2 | 5 | 6 | 7 | 9 | 1 | 4 | 8 |
| 6 | 7 | 9 | 4 | 1 | 8 | 3 | 5 | 2 |
| 1 | 9 | 4 | 8 | 6 | 7 | 5 | 2 | 3 |
| 2 | 5 | 6 | 9 | 3 | 1 | 4 | 8 | 7 |
| 8 | 3 | 7 | 5 | 4 | 2 | 9 | 6 | 1 |
| 5 | 4 | 3 | 7 | 8 | 6 | 2 | 1 | 9 |
| 9 | 6 | 2 | 1 | 5 | 3 | 8 | 7 | 4 |
| 7 | 1 | 8 | 2 | 9 | 4 | 6 | 3 | 5 |

Sudoku problem                          Sudoku solution

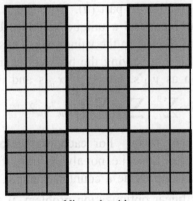

Nine sub-tables

**Figure 4.16**  Sudoku problem, solution, and sub-tables.

## Constraints

1. Each table entry contains exactly one digit.

$$\forall j, k \in \{1, \dots, 9\}, \sum_{i=1}^{9} x_{ijk} = 1$$

2. Each digit $i \in \{1, \dots, 9\}$ appears exactly once in each row.

   **First set of missing constraints (to be specified)**

3. Each digit $i \in \{1, \dots, 9\}$ appears exactly once in each column.

   **Second set of missing constraints (to be specified)**

4. Each digit $i \in \{1, ..., 9\}$ appears exactly once in each $3 \times 3$ sub-table.

$$\forall i \in \{1, ..., 9\}, \forall m, n \in \{1, 2, 3\}, \sum_{j=3m-2}^{3m} \sum_{k=3n-2}^{3n} x_{ijk} = 1$$

5. Some entries are already specified.

**Third set of missing constraints (to be specified)**

(*b*) Count the number of variables and the number of constraints in the completed problem formulation in part (*a*). Do not count the integer constraints.

(*c*) Suppose that Alice has solved a given Sudoku problem using the problem formulation in part (*a*). Now she is wondering whether the solved problem has a unique solution. Let $y_{ijk}$'s denote the solution found. Based on the values of $y_{ijk}$'s, specify an additional constraint such that solving the problem again will not yield the same solution as $y_{ijk}$'s. HINT: If $x_{ijk}$'s and $y_{ijk}$'s are exactly the same, what is $\sum_{i=1}^{9} \sum_{j=1}^{9} \sum_{k=1}^{9} y_{ijk} x_{ijk}$?

**Problem 4.7 (True or false)**  For each of the following statements, state whether it is true or false, i.e. not always true. If true, provide a brief justification. Otherwise, provide a counter-example.

(*a*) In an integer linear optimization problem, if the relaxed problem is feasible, then the original problem is also feasible.

(*b*) In an integer linear optimization problem, if the relaxed problem is infeasible, then the original problem is also infeasible.

(*c*) In an integer linear optimization problem, if the relaxed problem has an integer feasible solution as well as an optimal solution, then there is an optimal solution to the original problem.

(*d*) If an integer linear optimization problem can be solved to obtain an integer optimal solution through relaxation, then all the BFSs of the relaxed problem are integer solutions.

(*e*) If an integer linear optimization problem and its relaxed problem have the same optimal cost, then an integer optimal solution can always be obtained from solving the relaxed problem using the simplex algorithm.

(f) If an integer linear optimization problem contains finite lower and upper bounds for each variable, then there cannot be infinitely many optimal solutions.

(g) If a linear optimization problem contains finite lower and upper bounds for each variable, then there cannot be infinitely many optimal solutions.

# Reviews of Related Mathematics

This appendix provides brief reviews on basic mathematics used in the discussion of optimization theory in this book. The reviews include linear algebra and analysis. The reviews are not meant to be comprehensive, but are presented to help refresh relevant concepts. Results are stated without proofs. However, references are provided for more detailed information.

## A.1 Review of Linear Algebra

### Fields

A *field* $\mathcal{F}$ is a set of elements together with *addition* and *multiplication* defined to satisfy the following *field axioms*. The addition and multiplication of $a$ and $b$ are denoted by $a + b$ and $ab$ respectively.

**Axiom A.1 (Field axioms)**  For all $a, b, c \in \mathcal{F}$, the following properties hold:

1. *Commutativity*: $a + b = b + a$, $ab = ba$
2. *Associativity*: $(a + b) + c = a + (b + c)$, $(ab)c = a(bc)$
3. *Distributivity*: $a(b + c) = ab + ac$

4. *Existence of additive and multiplicative identities*: There exist elements, denoted by 0 and 1, such that $a + 0 = a$ and $1a = a$.

5. *Existence of additive inverse*: For each $a \in \mathcal{F}$, there exists an element, denoted by $-a$, such that $a + (-a) = 0$.

6. *Existence of multiplicative inverse*: For each $a \in \mathcal{F}$ and $a \neq 0$, there exists an element, denoted by $a^{-1}$, such that $aa^{-1} = 1$.

An example of a field is the set of *real numbers* $\mathbb{R}$ with the usual addition and multiplication. Another example of a field is the set of *complex numbers* $\mathbb{C}$ with complex addition and multiplication. Recall that, for $a, b \in \mathbb{C}$, the addition and multiplication of $a = a_R + ia_I$ and $b = b_R + ib_I$ are defined as $a + b = (a_R + b_R) + i(a_I + b_I)$ and $ab = (a_R b_R - a_I b_I) + i(a_R b_I + a_I b_R)$ respectively. For both $\mathbb{R}$ and $\mathbb{C}$, 0 is the additive identity while 1 is the multiplicative identity. This book only deals with the field $\mathbb{R}$.

## Vector Spaces

A *vector space* $\mathcal{V}$ is a set of elements defined over a field $\mathcal{F}$ according to the following *vector space* axioms. The elements of the field are called *scalars*. The elements of a vector space are called *vectors*.

**Axiom A.2 (Vector space axioms)** For all $\mathbf{u}, \mathbf{v}, \mathbf{w} \in \mathcal{V}$ and $\alpha, \beta \in \mathcal{F}$, the following properties hold:

1. *Closure*: $\mathbf{u} + \mathbf{v} \in \mathcal{V}$, $\alpha \mathbf{u} \in \mathcal{V}$

2. *Axioms for addition*

   Commutativity: $\mathbf{u} + \mathbf{v} = \mathbf{v} + \mathbf{u}$

   Associativity: $(\mathbf{u} + \mathbf{v}) + \mathbf{w} = \mathbf{u} + (\mathbf{v} + \mathbf{w})$

   Existence of identity: There exists an element in $\mathcal{V}$, denoted by $\mathbf{0}$, such that $\mathbf{u} + \mathbf{0} = \mathbf{u}$.

   Existence of inverse: For each $\mathbf{u} \in \mathcal{V}$, there exists an element in $\mathcal{V}$, denoted by $-\mathbf{u}$, such that $\mathbf{u} + (-\mathbf{u}) = \mathbf{0}$.

3. *Axioms for multiplication*

   Associativity: $(\alpha\beta)\mathbf{u} = \alpha(\beta\mathbf{u})$

   Unit multiplication: $1\mathbf{u} = \mathbf{u}$

   Distributivity: $\alpha(\mathbf{u} + \mathbf{v}) = \alpha\mathbf{u} + \alpha\mathbf{v}$, $(\alpha + \beta)\mathbf{u} + \alpha\mathbf{u} + \beta\mathbf{u}$

This book only deals with the scalar field $\mathbb{R}$. A vector space with the scalar field $\mathbb{R}$ is called a *real vector* space. In particular, this book focuses on the real vector space whose elements are $N$-dimensional real vectors in $\mathbb{R}^N$. For convenience, denote this vector space by $\mathbb{R}^N$.

A vector $\mathbf{v} \in \mathbb{R}^N$ can be described using $N$ real values as $\mathbf{v} = (v_1, \ldots, v_N)$. By convention, a vector $\mathbf{v} \in \mathbb{R}^N$ is considered as an $N \times 1$ matrix or a column vector.

A set of vectors $\mathbf{v}_1, \ldots, \mathbf{v}_M \in \mathcal{V}$ spans $\mathcal{V}$ if each $\mathbf{u} \in \mathcal{V}$ can be written as a *linear combination* of $\mathbf{v}_1, \ldots, \mathbf{v}_M$, i.e. $\mathbf{u} = \sum_{j=1}^{M} \alpha_j \mathbf{v}_j$ for some scalars $\alpha_1, \ldots, \alpha_M$. A vector space $\mathcal{V}$ is *finite-dimensional* if there is a finite set of vectors that spans $\mathcal{V}$.

A set of vectors $\mathbf{v}_1, \ldots, \mathbf{v}_M \in \mathcal{V}$ is *linearly dependent* if $\sum_{j=1}^{M} \alpha_j \mathbf{v}_j = 0$ for some scalars $\alpha_1, \ldots, \alpha_M$ not all equal to zero. A set of vectors $\mathbf{v}_1, \ldots, \mathbf{v}_M \in \mathcal{V}$ is *linearly independent* if it is not linearly dependent.

Consider an $M \times N$ matrix $\mathbf{A}$. Each element of $\mathbf{A}$ is referred to by the row index and the column index, i.e.

$$\mathbf{A} = \begin{bmatrix} a_{11} & a_{12} & \cdots & a_{1N} \\ a_{21} & a_{22} & \cdots & a_{2N} \\ \vdots & \vdots & \ddots & \vdots \\ a_{M1} & a_{M2} & \cdots & a_{MN} \end{bmatrix}$$

Column $j$ of $\mathbf{A}$, where $j \in \{1, \ldots, N\}$, is an $M$-dimensional column vector and will be denoted by $\mathbf{A}_j$. Row $i$ of $\mathbf{A}$, where $i \in \{1, \ldots, M\}$, is an $N$-dimensional row vector and will be denoted by $\mathbf{a}_i^{\mathrm{T}}$.[1]

The *rank* of $\mathbf{A}$, denoted by rank($\mathbf{A}$), is the dimension of the vector space spanned by the rows of $\mathbf{A}$. If $\mathbf{A}$ has rank $M$, i.e. all the rows of $\mathbf{A}$ are linearly independent, then $\mathbf{A}$ is said to be *full rank*.

---

[1]The notation "$\mathbf{a}$" is used to distinguish a row from a column. In addition, the notation "T" refers to the transpose operation and is used to emphasize the dimension of a row vector; recall that a vector is by convention a column vector.

## Bases and Dimensions

A set of vectors $\mathbf{v}_1, ..., \mathbf{v}_M \in \mathcal{V}$ is a *basis* for $\mathcal{V}$ if it spans $\mathcal{V}$ and is linearly independent. The following theorem states important properties of a basis [Gallager, 2008].

**Theorem A.1**  Let $\mathcal{V}$ be a finite-dimensional vector space.

1. If a set of vectors $\mathbf{v}_1, ..., \mathbf{v}_M \in \mathcal{V}$ spans $\mathcal{V}$ but is linearly dependent, then there is a subset of $\mathbf{v}_1, ..., \mathbf{v}_{M'}$ that forms a basis for $\mathcal{V}$ with $M' < M$ vectors.

2. If a set of vectors $\mathbf{v}_1, ..., \mathbf{v}_M \in \mathcal{V}$ is linearly independent but does not span $\mathcal{V}$, then there is a basis for $\mathcal{V}$ with $M' > M$ vectors that contains $\mathbf{v}_1, ..., \mathbf{v}_M$.

3. Every basis for $\mathcal{V}$ contains the same number of vectors.

Based on statement 3 of theorem A.1, the *dimension* of a finite-dimensional vector space is defined as the number of vectors in a basis. A vector space is *infinite-dimensional* if it is not finite-dimensional. For such a space, a basis must contain an infinite number of vectors. This book only deals with finite-dimensional vector spaces.

## Inner Product

An *inner product* defined on a vector space $\mathcal{V}$ (defined over a field $\mathcal{F}$) is a function of two vectors $\mathbf{u}, \mathbf{v} \in \mathcal{V}$, denoted by $\langle \mathbf{u}, \mathbf{v} \rangle$, that satisfies the following properties for all $\mathbf{u}, \mathbf{v}, \mathbf{w} \in \mathcal{V}$ and all $\alpha \in \mathcal{F}$.[2]

1. *Commutativity*: $\langle \mathbf{u}, \mathbf{v} \rangle = \langle \mathbf{v}, \mathbf{u} \rangle^*$

2. *Distributivity*: $\langle \mathbf{u} + \mathbf{v}, \mathbf{w} \rangle = \langle \mathbf{u}, \mathbf{w} \rangle + \langle \mathbf{v}, \mathbf{w} \rangle$

3. *Associativity*: $\langle \alpha\mathbf{u}, \mathbf{v} \rangle = \alpha\langle \mathbf{u}, \mathbf{v} \rangle$

4. Positivity: $\langle \mathbf{u}, \mathbf{u} \rangle \geq 0$ with equality if and only if $\mathbf{u} = \mathbf{0}$

Note that properties 1 and 2 imply that $\langle \mathbf{u}, \mathbf{v} + \mathbf{w} \rangle = \langle \mathbf{u}, \mathbf{v} \rangle + \langle \mathbf{u}, \mathbf{w} \rangle$, and properties 1 and 3 to show that $\langle \mathbf{u}, \alpha\mathbf{v} \rangle = \alpha^*\langle \mathbf{u}, \mathbf{v} \rangle$. A vector space with a defined inner product is called an *inner product space*.

In an inner product space $\mathcal{V}$, the *norm* of vector $\mathbf{u} \in \mathcal{V}$, denoted by $\|\mathbf{u}\|$, is defined as $\|\mathbf{u}\| = \sqrt{\langle \mathbf{u}, \mathbf{u} \rangle}$. Two vectors $\mathbf{u}, \mathbf{v} \in \mathcal{V}$ are called *orthogonal* if $\langle \mathbf{u}, \mathbf{v} \rangle = 0$.

---

[2] Let $x^*$ denote the complex conjugate of $x$.

This book focuses on the inner product space $\mathbb{R}^N$ consisting of all $N$-dimensional real vectors with the inner product of $\mathbf{u} = (u_1, ..., u_N)$ and $\mathbf{v} = (v_1, ..., v_N)$ defined as

$$\langle \mathbf{u}, \mathbf{v} \rangle = \sum_{i=1}^{N} u_i v_i. \qquad (A.1)$$

The corresponding norm is $\|\mathbf{u}\| = \sqrt{\sum_{i=1}^{N} u_i^2}$. The vector space $\mathbb{R}^N$ with the inner product defined in (A.1) is called the *N-dimensional Euclidean space*.

## Sub-spaces

A *sub-space* $S$ of a vector space $\mathcal{V}$ is a subset of $\mathcal{V}$ that is itself a vector space over the same scalar field.

For example, consider the Euclidean space $\mathbb{R}^3$. Let $\mathbf{u} = (1, 0, 0)$. Consider the set of all vectors of the form $\alpha\mathbf{u}$ for all $\alpha \in \mathbb{R}$. This set of vectors is itself a real vector space and is therefore a sub-space of $\mathbb{R}^3$. This sub-space has dimension 1 and has $\{\mathbf{u}\}$ as a basis.

Consider now the set of all vectors of the form $\alpha\mathbf{u}$ for all real $\alpha \neq 0$. This set of vectors is not a vector space since it does not contain the zero vector, and hence is not a sub-space of $\mathbb{R}^3$.

Let $\mathbf{v} = (1, 1, 0)$. Consider now the set of all vectors of the form $\alpha\mathbf{u} + \beta\mathbf{v}$ for all $\alpha, \beta \in \mathbb{R}$. This set of vectors is itself a real vector space and is therefore a sub-space of $\mathbb{R}^3$. Note that this sub-space has dimension 2 and has $\{\mathbf{u}, \mathbf{v}\}$ as a basis.

## Positive Semi-definite Matrices

An $N \times N$ symmetric matrix $\mathbf{A}$ is *positive semi-definite* if

$$\mathbf{x}^T\mathbf{A}\mathbf{x} \geq 0 \quad \text{for all } \mathbf{x} \in \mathbb{R}^N.$$

In addition, an $N \times N$ symmetric matrix $\mathbf{A}$ is *positive definite* if

$$\mathbf{x}^T\mathbf{A}\mathbf{x} > 0 \quad \text{for all nonzero } \mathbf{x} \in \mathbb{R}^N.$$

The fact that **A** is positive semi-definite is denoted by **A** $\geq$ 0. Similarly, the fact that **A** is positive definite is denoted by **A** > 0.[3]

Given matrix **A**, one method of testing whether **A** $\geq$ 0 is to compute the determinants of all the upperleft submatrices of **A**. If all these determinants are nonnegative, then **A** is positive semi-definite; otherwise, **A** is not positive semi-definite [Strang, 2005].[4] Similarly, for **A** > 0, all these determinants must be positive. For example, the determinant computations below imply that $\begin{bmatrix} 3 & 1 \\ 1 & 2 \end{bmatrix}$ is positive semi-definite while $\begin{bmatrix} 3 & 4 \\ 4 & 2 \end{bmatrix}$ is not.

$$\det([3]) = 3 \geq 0, \det\left(\begin{bmatrix} 3 & 1 \\ 1 & 2 \end{bmatrix}\right) = 5 \geq 0, \det\left(\begin{bmatrix} 3 & 4 \\ 4 & 2 \end{bmatrix}\right) = -10 < 0$$

## A.2 Review of Analysis

### Basic Definitions for Euclidean Spaces

A set $\mathcal{X}$ is a *metric space* if it is possible to define, for any two *points* $x, y \in \mathcal{X}$, an associated *metric* $d(x, y)$ with three properties.

---

[3]In this book, there are three meanings of the notation "$\geq$". The first meaning is for one real number to be greater than or equal to another, e.g. $4 \geq 2$. The second meaning is for one vector to have each of its component being greater than or equal to the same component of another vector, e.g. $\begin{bmatrix} 4 \\ 3 \end{bmatrix} \geq \begin{bmatrix} 1 \\ 2 \end{bmatrix}$ for $4 \geq 1$ and $3 \geq 2$. The third meaning is for positive semidefiniteness of a square matrix, e.g. $\begin{bmatrix} 1 & 0 \\ 0 & 2 \end{bmatrix} \geq 0$.

[4]Another method of checking positive semidefiniteness is based on the signs of the eigenvalues of **A**. If all eigenvalues are nonnegative, then **A** $\geq$ 0. In addition, if all eigenvalues are positive, then **A** > 0 [Strang, 2005].

1. *Non-negativity*: $d(x, y) \geq 0$ with equality if and only if $x = y$
2. *Symmetry*: $d(x, y) = d(y, x)$
3. *Triangle inequality*: $d(x, y) \leq d(x, z) + d(z, y)$ for any $z \in \mathcal{X}$

The metric space used throughout this book is the $N$-dimensional Euclidean space $\mathbb{R}^N$. Each point in this space is an $N$-dimensional vector in $\mathbb{R}^N$. For $\mathbf{x}, \mathbf{y} \in \mathbb{R}^N$, where $\mathbf{x} = (x_1, ..., x_N)$ and $\mathbf{y} = (y_1, ..., y_N)$, the metric $d(\mathbf{x}, \mathbf{y})$ is set equal to the *Euclidean distance*

$$d(\mathbf{x}, \mathbf{y}) = \sqrt{\sum_{i=1}^{N} (x_i - y_i)^2} \tag{A.2}$$

Recall that the norm of $\mathbf{x} \in \mathbb{R}^N$ is defined as $\|\mathbf{x}\| = \sqrt{\sum_{i=1}^{N} x_i^2}$. Accordingly, the Euclidean distance between $\mathbf{x}$ and $\mathbf{y}$ can also be written as $d(\mathbf{x}, \mathbf{y}) = \|\mathbf{x} - \mathbf{y}\|$.

For $r > 0$, a *radius-$r$ neighborhood* of $\mathbf{x}$, denoted by $\mathcal{N}_r(\mathbf{x})$, is defined as

$$\mathcal{N}_r(\mathbf{x}) = \{\mathbf{y} \in \mathbb{R}^N |\ \|\mathbf{y} - \mathbf{x}\| < r\}. \tag{A.3}$$

Note that $\mathcal{N}_r(\mathbf{x})$ is the $N$-sphere with radius $r$ centered at point $\mathbf{x}$. In $\mathbb{R}^2$, $\mathcal{N}_r(\mathbf{x})$ is the set of points inside the circle with radius $r$ centered at $\mathbf{x}$. In $\mathbb{R}$, $\mathcal{N}_r(\mathbf{x})$ is the open interval $(x - r, x + r)$.

Let $\mathcal{X}$ be a subset of $\mathbb{R}^N$. A point $\mathbf{p}$ is a *limit point* of $\mathcal{X}$ if every radius-$r$ neighborhood $\mathcal{N}_r(\mathbf{p})$ contains a point in $\mathcal{X}$ that is not $\mathbf{p}$. Note that a limit point of a set may not be a point in that set. For example, $p = 2$ is a limit point of the interval $(2, \infty)$ but is not in $(2, \infty)$.

A point $\mathbf{p}$ is an *interior point* of $\mathcal{X} \subset \mathbb{R}^N$ if there exists a neighborhood $\mathcal{N}_r(\mathbf{p})$ such that $\mathcal{N}_r(\mathbf{p}) \subset \mathcal{X}$. For example, $p = 2$ is not an interior point of $(2, \infty)$ while $p = 2.01$ is.

A set $\mathcal{X} \subset \mathbb{R}^N$ is *open* if every point in $\mathcal{X}$ is an interior point of $\mathcal{X}$. A set $\mathcal{X}$ is *closed* if its complement denoted by $\mathcal{X}^c$ is open.[5] For example, $(2, \infty)$ is open while $(-\infty, 2]$ is closed. Note that some sets, e.g. $\mathbb{R}$ and $\emptyset$, are both open and closed.[6]

---

[5] An equivalent definition of a closed set is that a set $\mathcal{X}$ is closed if it contains all of its limit points.

[6] Therefore, the statement "$\mathcal{X}$ is closed if it is not open" is wrong.

The *closure* of a set $\mathcal{X} \subset \mathbb{R}^N$, denoted by $\bar{\mathcal{X}}$, contains all the points of $\mathcal{X}$ together with all the limit points of $\mathcal{X}$. From the definition of a closed set, it is clear that the closure of any set is a closed set.

A set $\mathcal{X} \subset \mathbb{R}^N$ is *bounded* if there exists some $M > 0$ such that $\|\mathbf{x}\| < M$ for all $\mathbf{x} \in \mathcal{X}$. A set $\mathcal{X} \subset \mathbb{R}^N$ that is closed and bounded is called *compact*.

## Sequences and Limits

A *sequence* $\mathbf{x}_1, \mathbf{x}_2, \ldots$ is typically denoted by $\{\mathbf{x}_i\}$ or simply $\mathbf{x}_i$. A sequence $\mathbf{x}_i$ in $\mathcal{X} \subset \mathbb{R}^N$ *converges* if there is a limit point $\mathbf{p} \in \mathcal{X}$ such that, for any $\varepsilon > 0$, there exists a positive integer $N$ such that $\|\mathbf{x}_i - \mathbf{p}\| < \varepsilon$ for all $i \geq N$. The fact that $\mathbf{x}_i$ converges to $\mathbf{p}$ is denoted by

$$\mathbf{x}_i \to \mathbf{p} \quad \text{or} \quad \lim_{i \to \infty} \mathbf{x}_i = \mathbf{p}.$$

Note that the limit of a convergent sequence must be in the set $\mathcal{X}$ to which the sequence belongs. For example, the sequence 1, 1/2, 1/4, 1/8, ... converges in the set $[0, \infty)$ but *diverges* (i.e. does not converge) in $(0, \infty)$.

It is worth noting that if $\mathbf{p} \in \mathcal{X}$ is a limit point of $\mathcal{X}$, then there exists a sequence $\mathbf{x}_i$ in $\mathcal{X}$ that converges to $\mathbf{p}$. This sequence can be constructed simply by setting $\mathbf{x}_i$ equal to any point in $\mathcal{N}_r^i(\mathbf{p}) \cap \mathcal{X}$, where $r_i$ converges to 0, e.g. $r_i = 2^{-i}$.

## Functions and Limits

A *function* is a mapping from a point in a set called the *domain* to a point in a set called the range. This book deals with a *real-valued* function $f$ that maps a point in a subset $\mathcal{X} \subset \mathbb{R}^N$ to a real value. Such a function is denoted as $f\colon \mathcal{X} \to \mathbb{R}$.

Consider a function $f\colon \mathcal{X} \to \mathbb{R}$. Let $\mathbf{p}$ be a limit point of $\mathcal{X}$. If there is a point $q \in \mathbb{R}$ such that, for any $\varepsilon > 0$, there exists $\delta > 0$ such that

$$|f(\mathbf{x}) - q| < \varepsilon \text{ for all } \mathbf{x} \in \mathcal{X} \text{ with } 0 < \|\mathbf{x} - \mathbf{p}\| < \delta,$$

then $q$ is said to be the *limit* of $f(\mathbf{x})$ *as* $\mathbf{x}$ *approaches* $\mathbf{p}$; this fact is denoted by

$$\lim_{\mathbf{x} \to \mathbf{p}} f(\mathbf{x}) = q.$$

Note that $\lim_{\mathbf{x} \to \mathbf{p}} f(\mathbf{x}) = q$ implies that, for any sequence $\mathbf{x}_i$ that converges to $\mathbf{p}$, $\lim_{i \to \infty} f(\mathbf{x}_i) = q$.

## Continuity

Let $\mathbf{p}$ be a limit point of $\mathcal{X} \subset \mathbb{R}^N$. A function $f: \mathcal{X} \to \mathbb{R}$ is *continuous at* $\mathbf{p} \in \mathcal{X}$ if $\lim_{\mathbf{x} \to \mathbf{p}} f(\mathbf{x}) = f(\mathbf{p})$. A function $f$ is *continuous on* $\mathcal{X}$ if it is continuous at every point in $\mathcal{X}$. The following theorem is a consequence of continuity. Its proof can be found in [Rudin, 1976].

**Theorem A.2 (Intermediate value theorem)**    Consider a real function $f: \mathcal{X} \to \mathbb{R}$. Let $\mathbf{x}, \mathbf{x} + \mathbf{d} \in \mathcal{X}$. Suppose that $f$ is continuous at every point of the form $\mathbf{x} + \alpha\mathbf{d}$ for $\alpha \in [0, 1]$, i.e. $f$ is continuous on the line segment between $\mathbf{x}$ and $\mathbf{x} + \mathbf{d}$. If $f(\mathbf{x}) < c < f(\mathbf{x} + \mathbf{d})$, then there exists a $\beta \in [0, 1]$ such that $f(\mathbf{x} + \beta\mathbf{d}) = c$.

Consider a function $f: \mathcal{X} \to \mathbb{R}$. When $\mathcal{X}$ is a compact subset of $\mathbb{R}^N$, there exists a *maximum* point as well as a *minimum* point in $\mathcal{X}$. This result is also known as the *Weierstrass theorem*. Its proof can be found in [Rudin, 1976].

**Theorem A.3 (Weierstrass theorem)**    Consider a real function $f: \mathcal{X} \to \mathbb{R}$. Suppose that $f$ is continuous on $\mathcal{X}$ and that $\mathcal{X}$ is a compact subset of $\mathbb{R}^N$. Then, there exists a maximum point and a minimum point of $f$ in $\mathcal{X}$. More specifically, there are points $\mathbf{x}_{max}, \mathbf{x}_{min} \in \mathcal{X}$ such that

$$f(\mathbf{x}_{max}) \geq f(\mathbf{x}) \quad \text{for all } \mathbf{x} \in \mathcal{X},$$

$$f(\mathbf{x}_{min}) \leq f(\mathbf{x}) \quad \text{for all } \mathbf{x} \in \mathcal{X}.$$

To illustrate the need for compactness in the Weierstrass theorem, consider $f: [0, 1) \to \mathbb{R}$ defined as $f(x) = x$. Note that $[0, 1)$ is not compact. In addition, there is no maximum point in $[0, 1)$.

## Differentiation

Let $\mathcal{X} \subset \mathbb{R}^N$. A function $f: \mathcal{X} \to \mathbb{R}$ is *differentiable at* $\mathbf{x} \in \mathcal{X}$ if (1) for all $i \in \{1, \ldots, N\}$, there exists the limit

$$\frac{\partial f(\mathbf{x})}{\partial x_i} = \lim_{\alpha \to 0} \frac{f(\mathbf{x} + \alpha \mathbf{e}_i) - f(\mathbf{x})}{\alpha},$$

where $\mathbf{e}_i$ is the unit vector with a "1" in the $i$th position, and (2) each $\partial f(\mathbf{x})/\partial x_i$ is continuous in some neighborhood of $\mathbf{x}$. The quantity $\partial f(\mathbf{x})/\partial x_i$ is called the *partial derivative* of $f$ with respect to $x_i$. The *gradient* of $f$, denoted by $\nabla f$, is defined as the vector

$$\nabla f(\mathbf{x}) = \begin{bmatrix} \partial f(\mathbf{x}) / \partial x_1 \\ \vdots \\ \partial f(\mathbf{x}) / \partial x_N \end{bmatrix} \tag{A.4}$$

It should be noted that $f$ is differentiable at $\mathbf{x} \in \mathcal{X}$ only if it is continuous at $\mathbf{x}$. A function $f$ is *differentiable on* $\mathcal{X}$ if it is differentiable at every point in $\mathcal{X}$.

If each partial derivative of $f$ is differentiable on $\mathcal{X}$, then $f$ is called *twice differentiable* on $\mathcal{X}$. In this case, the *Hessian* of $f$, denoted by $\nabla^2 f$, is defined as the matrix

$$\nabla^2 f(\mathbf{x}) = \begin{bmatrix} \dfrac{\partial^2 f(\mathbf{x})}{\partial x_1^2} & \cdots & \dfrac{\partial^2 f(\mathbf{x})}{\partial x_1 \partial x_N} \\ \vdots & \ddots & \vdots \\ \dfrac{\partial^2 f(\mathbf{x})}{\partial x_N \partial x_1} & \cdots & \dfrac{\partial^2 f(\mathbf{x})}{\partial x_N^2} \end{bmatrix} \tag{A.5}$$

where $\dfrac{\partial^2 f(\mathbf{x})}{\partial x_i \partial x_j}$ is the partial derivative of $\dfrac{\partial f(\mathbf{x})}{\partial x_i}$ with respect to $x_j$.

The following theorems involve derivatives of functions. Their proofs can be found in [Apostol, 1969].

**Theorem A.4 (Mean value theorem)** Consider a real function $f \colon \mathcal{X} \to \mathbb{R}$. Let $\mathbf{x}, \mathbf{x} + \mathbf{d} \in \mathcal{X}$. Suppose that $f$ is differentiable at every point of the form $\mathbf{x} + \alpha \mathbf{d}$ for $\alpha \in [0, 1]$, i.e. $f$ is differentiable on the line segment between $\mathbf{x}$ and $\mathbf{x} + \mathbf{d}$. Then, there exists a $\beta \in [0, 1]$ such that

$$f(\mathbf{x} + \mathbf{d}) = f(\mathbf{x}) + \nabla f(\mathbf{x} + \beta \mathbf{d})^\mathrm{T} \mathbf{d}.$$

**Theorem A.5 (Second-order Taylor series expansion)** Consider a real function $f: \mathcal{X} \to \mathbb{R}$. Let $\mathbf{x} \in \mathcal{X}$. Suppose that $f$ is twice differentiable on the neighborhood $\mathcal{N}_r(\mathbf{x})$. Let $\mathbf{x} + \mathbf{d} \in \mathcal{N}_r(\mathbf{x})$. Then, there exists a $\beta \in [0, 1]$ such that

$$f(\mathbf{x} + \mathbf{d}) = f(\mathbf{x}) + \nabla f(\mathbf{x})^{\mathrm{T}}\mathbf{d} + \frac{1}{2}\mathbf{d}^{\mathrm{T}}\nabla^2 f(\mathbf{x} + \beta\mathbf{d})\mathbf{d}.$$

For a vector $\mathbf{d} \in \mathbb{R}^N$ with small $\|\mathbf{d}\|$, the mean value theorem yields the *first-order Taylor series approximation* shown below.

$$f(\mathbf{x} + \mathbf{d}) \approx f(\mathbf{x}) + \nabla f(\mathbf{x})^{\mathrm{T}}\mathbf{d} \tag{A.6}$$

In addition, for a vector $\mathbf{d} \in \mathbb{R}^N$ with small $\|\mathbf{d}\|$, theorem A.5 yields the *second-order Taylor series approximation* shown below.

$$f(\mathbf{x} + \mathbf{d}) \approx f(\mathbf{x}) + \nabla f(\mathbf{x})^{\mathrm{T}}\mathbf{d} + \frac{1}{2}\mathbf{d}^{\mathrm{T}} \nabla^2 f(\mathbf{x})\mathbf{d} \tag{A.7}$$

# Solutions to Exercise Problems

## Solutions for Chapter 2

**Solution 2.1 (Convexity of a set):**

(a) The given set $\mathcal{X}$ is not convex. To see why, consider two points $\mathbf{x} = (a, b + 2)$ and $\mathbf{y} = (a + 2, b)$, which are both in $\mathcal{X}$. Their convex combination $\mathbf{x}/2 + \mathbf{y}/2 = (a + 1, b + 1)$ is not in $\mathcal{X}$.

(b) The given hyperplane $\mathcal{X}$ is convex. To see why, consider any convex combination $\mathbf{z} = \alpha\mathbf{x} + (1 - \alpha)\mathbf{y}$, where $\mathbf{x}, \mathbf{y} \in \mathcal{X}$, $\mathbf{x} \neq \mathbf{y}$, and $\alpha \in [0, 1]$. Since $\mathbf{a}^T\mathbf{x} = \mathbf{a}^T\mathbf{y} = b$,

$$\mathbf{a}^T\mathbf{z} = \alpha\mathbf{a}^T\mathbf{x} + (1 - \alpha)\mathbf{a}^T\mathbf{y} = \alpha b + (1 - \alpha)b = b,$$

which implies that $\mathbf{z} \in \mathcal{X}$.

(c) The given halfspace $\mathcal{X}$ is convex. To see why, consider any convex combination $\mathbf{z} = \alpha\mathbf{x} + (1 - \alpha)\mathbf{y}$, where $\mathbf{x}, \mathbf{y} \in \mathcal{X}$, $\mathbf{x} \neq \mathbf{y}$, and $\alpha \in [0, 1]$. Since $\mathbf{a}^T\mathbf{x} \geq b$ and $\mathbf{a}^T\mathbf{y} \geq b$,

$$\mathbf{a}^T\mathbf{z} = \alpha\mathbf{a}^T\mathbf{x} + (1 - \alpha)\mathbf{a}^T\mathbf{y} \geq \alpha b + (1 - \alpha)b = b,$$

which implies that $\mathbf{z} \in \mathcal{X}$.

(d) The given polyhedron $\mathcal{X}$ is convex. To see why, consider any convex combination $\mathbf{z} = \alpha\mathbf{x} + (1 - \alpha)\mathbf{y}$, where $\mathbf{x}, \mathbf{y} \in \mathcal{X}, \mathbf{x} \neq \mathbf{y}$, and $\alpha \in [0, 1]$. Since $\mathbf{Ax} \geq \mathbf{b}$, $\mathbf{Ay} \geq \mathbf{b}$, $\mathbf{Cx} = \mathbf{d}$, and $\mathbf{Cy} = \mathbf{d}$,

$$\mathbf{Az} = \alpha\mathbf{Ax} + (1 - \alpha)\mathbf{Ay} \geq \alpha\mathbf{b} + (1 - \alpha)\mathbf{b} = \mathbf{b},$$
$$\mathbf{Cz} = \alpha\mathbf{Cx} + (1 - \alpha)\mathbf{Cy} = \alpha\mathbf{d} + (1 - \alpha)\mathbf{d} = \mathbf{d},$$

which implies that $\mathbf{z} \in \mathcal{X}$.

(e) The given $N$-dimensional ball $\mathcal{X}$ is convex. To see why, consider any convex combination $\mathbf{z} = \alpha\mathbf{x} + (1 - \alpha)\mathbf{y}$, where $\mathbf{x}, \mathbf{y} \in \mathcal{X}, \mathbf{x} \neq \mathbf{y}$, and $\alpha \in [0, 1]$. Since $\|\mathbf{x} - \mathbf{a}\| \leq r$, $\|\mathbf{y} - \mathbf{a}\| \leq r$, and $\|\mathbf{p} + \mathbf{q}\| \leq \|\mathbf{p}\| + \|\mathbf{q}\|$ for any real vectors $\mathbf{p}, \mathbf{q} \in \mathbb{R}^N$,

$$\|\mathbf{z} - \mathbf{a}\| = \|\alpha(\mathbf{x} - \mathbf{a}) + (1 - \alpha)(\mathbf{y} - \mathbf{a})\|$$
$$\leq \alpha\|\mathbf{x} - \mathbf{a}\| + (1 - \alpha)\|\mathbf{y} - \mathbf{a}\|$$
$$\leq \alpha r + (1 - \alpha)r = r,$$

which implies that $\mathbf{z} \in \mathcal{X}$.

(f) The given convex hull $\mathcal{X}$ is convex. To see why, consider any convex combination $\mathbf{z} = \alpha\mathbf{x} + (1 - \alpha)\mathbf{y}$, where $\mathbf{x}, \mathbf{y} \in \mathcal{X}, \mathbf{x} \neq \mathbf{y}$, and $\alpha \in [0, 1]$. Since $\mathbf{x}, \mathbf{y} \in \mathcal{X}$, there exist coefficients $a_1, \ldots, a_M$ and $b_1, \ldots, b_M$ such that

$$\mathbf{x} = a_1\mathbf{v}_1 + \cdots + a_M\mathbf{v}_M, \quad \sum_{m=1}^{M} a_m = 1, a_m \geq 0 \quad \text{for all } m,$$

$$\mathbf{y} = b_1\mathbf{v}_1 + \cdots + b_M\mathbf{v}_M, \quad \sum_{m=1}^{M} b_m = 1, b_m \geq 0 \quad \text{for all } m.$$

Consequently,

$$\mathbf{z} = \alpha\mathbf{x} + (1 - \alpha)\mathbf{y} = (\alpha a_1 + (1 - \alpha)b_1)\mathbf{v}_1 + \cdots$$
$$+ (\alpha a_M + (1 - \alpha)b_M)\mathbf{v}_M = c_1\mathbf{v}_1 + \cdots + c_M\mathbf{v}_M,$$

where $c_m = \alpha a_m + (1 - \alpha)b_m$ for all $m$. In addition,

$$\sum_{m=1}^{M} c_m = \alpha\sum_{m=1}^{M} a_m + (1 - \alpha)\sum_{m=1}^{M} b_m = \alpha\cdot1 + (1 - \alpha)\cdot1 = 1,$$

$$c_m = \alpha a_m + (1 - \alpha)b_m \geq \alpha \cdot 0 + (1 - \alpha) \cdot 0 = 0.$$

It follows that $\mathbf{z} \in \mathcal{X}$.

(g) The given intersection $\mathcal{X} = \bigcap_{m=1}^{M} \mathcal{A}_m$ is convex. To see why, consider any convex combination $\mathbf{z} = \alpha \mathbf{x} + (1 - \alpha)\mathbf{y}$, where $\mathbf{x}$, $\mathbf{y} \in \mathcal{X}$, $\mathbf{x} \neq \mathbf{y}$, and $\alpha \in [0, 1]$. For all $m$, since $\mathbf{x}, \mathbf{y} \in \mathcal{A}_m$, $\mathbf{z} \in \mathcal{A}_m$ by the convexity of $\mathcal{A}_m$. It follows that $\mathbf{z} \in \bigcap_{m=1}^{M} \mathcal{A}_m$.

(h) The given union $\mathcal{X} = \bigcup_{m=1}^{M} \mathcal{A}_m$ may not be convex in general. For a counterexample, consider two halfspaces $\{\mathbf{x} \in \mathbb{R}^2 | \, x_1 \leq a\}$ and $\{\mathbf{x} \in \mathbb{R}^2 | \, x_2 \leq b\}$, where $a, b \in \mathbb{R}$. Both halfspaces are convex sets. However, their union results in the set of part (a), which is not convex.    □

## Solution 2.2 (Convexity of a function):

(a) For $\mathcal{X} = \mathbb{R}$, $f(x) = x^3$ is not convex. In particular, the second-order derivative $d^2 f(x)/dx^2 = 6x$ is negative for $x < 0$, violating the second-order condition in theorem 2.2.

(b) For $\mathcal{X} = \mathbb{R}^+$, $f(x) = x^3$ is convex. In particular, the second-order derivative $d^2 f(x)/dx^2 = 6x$ is nonnegative for $x \in \mathbb{R}^+$, satisfying the second-order condition in theorem 2.2.

(c) The given function $f$ is not convex. In particular, $\nabla^2 f(\mathbf{x}) = \mathbf{A}$ is not positive semi-definite since $\det(\mathbf{A}) = -4$. Hence, the second-order condition in theorem 2.2 is violated.

(d) The given function $f$ is convex. In particular, $\nabla^2 f(\mathbf{x}) = \mathbf{A}$ is positive semi-definite since $\det([4]) = 4$ and $\det(\mathbf{A}) = 11$. Hence, the second-order condition in theorem 2.2 is satisfied.

(e) The minimum $f$ of two linear functions may not be convex in general. For a counterexample, consider $f(x) = \min(x, -x)$ and two points $-1$ and $1$. Since

$$0 = f\left(\frac{1}{2} \cdot -1 + \frac{1}{2} \cdot 1\right) > \frac{1}{2} f(-1) + \frac{1}{2} f(1) = -1,$$

the function $f$ is not convex.

(*f*) The maximum $f$ of two linear functions is convex. To see why, consider any convex combination $\mathbf{z} = \alpha\mathbf{x} + (1 - \alpha)\mathbf{y}$, where $\mathbf{x}$, $\mathbf{y} \in \mathcal{X}$, $\mathbf{x} \neq \mathbf{y}$, and $\alpha \in [0, 1]$. Since $\max_{\mathbf{u}} (p(\mathbf{u}) + q(\mathbf{u})) \leq \max_{\mathbf{u}} p(\mathbf{u}) + \max_{\mathbf{u}} q(\mathbf{u})$ for any real functions $p$ and $q$,

$$f(\mathbf{z}) = \max(\alpha\mathbf{a}^T\mathbf{x} + (1 - \alpha)\mathbf{a}^T\mathbf{y}, \ \alpha\mathbf{b}^T\mathbf{x} + (1 - \alpha)\mathbf{b}^T\mathbf{y})$$

$$= \max_{\mathbf{u} \in \{\mathbf{a}, \mathbf{b}\}} (\alpha\mathbf{u}^T\mathbf{x} + (1 - \alpha)\mathbf{u}^T\mathbf{y})$$

$$\leq \alpha \max_{\mathbf{u} \in \{\mathbf{a}, \mathbf{b}\}} (\mathbf{u}^T\mathbf{x}) + (1 - \alpha) \max_{\mathbf{u} \in \{\mathbf{a}, \mathbf{b}\}} (\mathbf{u}^T\mathbf{y})$$

$$= \alpha f(\mathbf{x}) + (1 - \alpha) f(\mathbf{y}),$$

which implies that $f$ is convex.                                              □

**Solution 2.3 (Convexity of an optimization problem)**   Since the cost function $f$ is assumed to be convex, it remains to show that the feasible set $\mathcal{F} = \{\mathbf{x} \in \mathcal{X}|g(\mathbf{x}) \leq 0\}$ is convex. To do so, consider any convex combination $\mathbf{z} = \alpha\mathbf{x} + (1 - \alpha)\mathbf{y}$, where $\mathbf{x}, \mathbf{y} \in \mathcal{X}$, $\mathbf{x} \neq \mathbf{y}$, and $\alpha \in [0, 1]$. Note that

$$g(\mathbf{z}) \leq \alpha g(\mathbf{x}) + (1 - \alpha) g(\mathbf{y}) \leq 0,$$

where the first inequality follows from the convexity of $g$, while the last inequality follows from the fact that $g(\mathbf{x}) \leq 0$ and $g(\mathbf{y}) \leq 0$. It follows that $\mathbf{z} \in \mathcal{F}$, implying the convexity of $\mathcal{F}$.                                              □

**Solution 2.4 (Convexity of an optimization problem)**   First, the feasible set $\mathcal{F} = \left\{ x \in \mathbb{R}^N \middle| \sum_{i=1}^{N} x_i = 1, \mathbf{x} \geq \mathbf{0} \right\}$ is shown to be convex as follows. Consider any convex combination $\mathbf{z} = \alpha\mathbf{x} + (1 - \alpha)\mathbf{y}$, where $\mathbf{x}$, $\mathbf{y} \in \mathcal{X}$, $\mathbf{x} \neq \mathbf{y}$, and $\alpha \in [0, 1]$. Since $\mathbf{x}, \mathbf{y} \in \mathcal{F}$, $\sum_{i=1}^{N} x_i = \sum_{i=1}^{N} y_i = 1$, $\mathbf{x} \geq \mathbf{0}$, and $\mathbf{y} \geq \mathbf{0}$. Consequently,

$$\sum_{i=1}^{N} z_i = \alpha \sum_{i=1}^{N} x_i + (1 - \alpha) \sum_{i=1}^{N} y_i = 1,$$

$$\mathbf{z} = \alpha\mathbf{x} + (1 - \alpha)\mathbf{y} \geq \alpha\mathbf{0} + (1 - \alpha)\mathbf{0} = \mathbf{0}.$$

It follows that $\mathbf{z} \in \mathcal{F}$, implying the convexity of $\mathcal{F}$.

Let $f(\mathbf{x}) = \sum_{i=1}^{N} x_i \ln x_i$. The Hessian of $f$ is equal to the diagonal matrix

$$\nabla^2 f(\mathbf{x}) = \begin{bmatrix} 1/x_1 & 0 & \cdots & 0 \\ 0 & 1/x_2 & \cdots & 0 \\ \vdots & \vdots & \ddots & \vdots \\ 0 & 0 & \cdots & 1/x_N \end{bmatrix},$$

which is positive semi-definite for $\mathbf{x} \geq \mathbf{0}$. Using the second-order condition in theorem 2.2, it follows that $f$ is convex. In conclusion, the given problem is a convex optimization problem.   □

**Solution 2.5 (Using the KKT conditions):**

(a) Let $f(x) = -xe^{-x}$. The second derivative of $f$ is equal to

$$\frac{d^2 f(x)}{dx^2} = (2 - x)e^{-x},$$

which is positive for $x \leq s < 2$. Hence, the second-order condition in theorem 2.2 is satisfied, implying that $f$ is convex in the feasible set.

(b) The Lagrangian is

$$\Lambda(x, \lambda) = -xe^{-x} + \lambda(x - s).$$

From Lagrangian optimality,

$$0 = \frac{\partial \Lambda(x^*, \lambda^*)}{\partial x} = (x^* - 1)e^{-x^*} + \lambda^*$$

$$\Rightarrow \qquad \lambda^* = (1 - x*)e^{-x}.$$

If $\lambda^* > 0$, then $x^* = s$ from complimentary slackness. Since $\lambda^* = (1 - x^*)e^{-x^*}$, it follows that $\lambda^* > 0$ is only possible if $s < 1$. Hence, we consider two cases.

1. $s \geq 1$: In this case, $\lambda^* = 0$ from the above argument. From $\lambda^* = (1 - x^*)e^{-x}$, it follows that $x^* = 1$.

2. $s < 1$: If $\lambda^* > 0$, then $x^* = s$, yielding $\lambda^* = (1 - s)e^{-s}$. If $x^* > s$, then $\lambda^* = 0$, yielding $x^* = 1$, which is a contradiction since $x^* \le s < 1$. In summary, the primal-dual optimal solution pair is

$$(x^*, \lambda^*) = \begin{cases} (1, 0) & s \ge 1 \\ (s, (1-s)e^{-s}) & s < 1 \end{cases}$$

Note that the solution depends on the value of $s$.

**Solution 2.6 (Using the KKT conditions)**                                    □

(a) Since the cost function is linear and thus convex, it remains to show that the feasible set $\mathcal{F} = \left\{ \mathbf{x} \in \mathbb{R}^N \middle| \sum_{i=1}^{N} e^{-x_i} \le 1 \right\}$ is convex. Consider any convex combination $\mathbf{z} = \alpha\mathbf{x} + (1 - \alpha)\mathbf{y}$, where $\mathbf{x}$, $\mathbf{y} \in \mathcal{X}$, $\mathbf{x} \ne \mathbf{y}$, and $\alpha \in [0, 1]$. Since $\mathbf{x}$, $\mathbf{y} \in \mathcal{F}$, $\sum_{i=1}^{N} e^{-x_i} \le 1$ and $\sum_{i=1}^{N} e^{-y_i} \le 1$. Consequently,

$$\sum_{i=1}^{N} e^{-z_i} \le \alpha \sum_{i=1}^{N} e^{-x_i} + (1 - \alpha) \sum_{i=1}^{N} e^{-y_i}$$

$$\le \alpha \cdot 1 + (1 - \alpha) \cdot 1 = 1,$$

where the first inequality follows from the convexity of function $f(u) = e^{-u}$. It follows that $\mathbf{z} \in \mathcal{F}$, implying the convexity of $\mathcal{F}$.

(b) The Lagrangian is

$$\Lambda(\mathbf{x}, \lambda) = \sum_{i=1}^{N} c_i x_i + \lambda \left( \sum_{i=1}^{N} e^{-x_i} - 1 \right).$$

Recall that the dual function is $q(\lambda) = \inf_{\mathbf{x} \in \mathbb{R}^N} \Lambda(\mathbf{x}, \lambda)$. In this problem, it is possible to minimize $\Lambda(\mathbf{x}, \lambda)$ to get the dual function as follows.

$$0 = \frac{\partial \Lambda(\mathbf{x}^*, \lambda)}{\partial x_i} = c_i - \lambda e^{-x_i^*} \Rightarrow x_i^* = \ln \frac{\lambda}{c_i}$$

Substituting $x_i^* = \ln(\lambda/c_i)$ in the Lagrangian expression yields the dual function below.

$$q(\lambda) = 1 - \lambda + \sum_{i=1}^{N} c_i \ln \frac{\lambda}{c_i}$$

The corresponding dual problem is as follows.

$$\text{maximize } 1 - \lambda + \sum_{i=1}^{N} c_i \ln \frac{\lambda}{c_i}$$

$$\text{subject to } \lambda \geq 0$$

(c) From Lagrangian optimality,

$$0 = \frac{\partial \Lambda(\mathbf{x}^*, \lambda^*)}{\partial x_i} = c_i - \lambda^* e^{-x_i^*} \Rightarrow x_i^* = \ln \frac{\lambda^*}{c_i}.$$

From $c_i - \lambda^* e^{-x_i^*} = 0$ and the assumption that $c_i > 0$ for all $i$, it is not possible to have $\lambda^* = 0$. Thus, $\lambda^* > 0$. Complimentary slackness then implies that

$$1 = \sum_{i=1}^{N} e^{-x_i^*} = \sum_{i=1}^{N} \frac{c_i}{\lambda^*} \Rightarrow \lambda^* \sum_{i=1}^{N} c_i = 1.$$

It follows that $x_i^* = -\ln c_i$. In summary, the primal-dual optimal solution pair is $(x_1^*, \ldots, x_N^*, \lambda^*) = (-\ln c_1, \ldots, -\ln c_N, 1)$. □

**Solution 2.7 (Convexification of an optimization problem):**

(a) No, the problem is not always a convex optimization problem. As a counterexample, consider $a = b = 3$. Consequently, the Hessian of the cost function $f(x_1, x_2) = -x_1^3 x_2^3$ is

$$\nabla^2 f(x_1, x_2) = \begin{bmatrix} -6x_1 x_2^3 & -9x_1^2 x_2^2 \\ -9x_1^2 x_2^2 & -6x_1^3 x_2 \end{bmatrix}.$$

Since $\det([-6x_1 x_2^3]) < 0$ for $x_1, x_2 > 0$, $\nabla^2 f$ is not positive semi-definite, implying that the objective function $f$ is not convex.

(b) From the changes of variables, $x_1 = e^{y_1}$ and $x_2 = e^{y_2}$. It follows that the optimization problem in part (a) can be written in terms of $y_1$ and $y_2$ as follows

$$\text{minimize} \quad -e^{ay_1 + by_2}$$

$$\text{subject to} \quad e^{y_1} + e^{y_2} \leq 1$$

$$e^{y_1}, \; e^{y_2} \geq 0$$

Note that the non-negativity constraints can be omitted since $e^u \geq 0$ for all $u$. For the objective function, minimizing $-e^{ay_1 + by_2}$ is equivalent to maximizing $ay_1 + by_2$, which is in turn equivalent to minimizing $-ay_1 - by_2$.

(c) Since the cost function is linear, it is convex. It remains to show that the feasible set $\mathcal{F} = \{(y_1, y_2) \in \mathbb{R}^2 | e^{y_1} + e^{y_2} \leq 1\}$ is convex. Consider any convex combination $\mathbf{z} = \alpha \mathbf{x} + (1 - \alpha)\mathbf{y}$, where $\mathbf{x}$, $\mathbf{y} \in \mathcal{F}$, $\mathbf{x} \neq \mathbf{y}$, and $\alpha \in [0, 1]$. From the convexity of the function $e^u$ together with the facts that $e^{x_1} + e^{x_2} \leq 1$ and $e^{y_1} + e^{y_2} \leq 1$,

$$e^{z_1} + e^{z_2} = e^{\alpha x_1 + (1 - \alpha)x_2} + e^{\alpha y_1 + (1 - \alpha)y_2}$$

$$\leq \alpha e^{x_1} + (1 - \alpha)e^{y_1} + \alpha e^{x_2} + (1 - \alpha)e^{y_2}$$

$$= \alpha (e^{x_1} + e^{x_2}) + (1 - \alpha)(e^{y_1} + e^{y_2})$$

$$\leq \alpha \cdot 1 + (1 - \alpha) \cdot 1 = 1,$$

which implies that $\mathbf{z} \in \mathcal{F}$.

(d) The Lagrangian for the problem is

$$\Lambda(\mathbf{y}, \lambda) = -ay_1 - by_2 + \lambda(e^{y_1} + e^{y_2} - 1).$$

The dual function is $q(\lambda) = \inf_{\mathbf{y} \in \mathbb{R}^2} \Lambda(\mathbf{y}, \lambda)$, and can be obtained in this case by solving for $\mathbf{y}^*$ such that $\partial \Lambda(\mathbf{y}^*, \lambda)/dy_1 = \partial \Lambda(\mathbf{y}^*, \lambda)/dy_2 = 0$, yielding

$$\frac{\partial \Lambda(\mathbf{y}^*, \lambda)}{\partial y_1} = -a + \lambda e^{y_1^*} = 0 \Rightarrow y_1^* = \ln \frac{a}{\lambda},$$

$$\frac{\partial \Lambda(\mathbf{y}^*, \lambda)}{\partial y_2} = -b + \lambda e^{y_2^*} = 0 \Rightarrow y_2^* = \ln \frac{b}{\lambda},$$

for $\lambda > 0$. In addition, note that $q(\lambda) = -\infty$ for $\lambda \le 0$. The corresponding dual function is

$$q(\lambda) = \begin{cases} -a \ln \dfrac{a}{\lambda} - b \ln \dfrac{b}{\lambda} + (a+b-1), & \lambda > 0 \\ -\infty & \lambda \le 0 \end{cases}$$

yielding the dual problem as shown below.

maximize $-a \ln \dfrac{a}{\lambda} - b \ln \dfrac{b}{\lambda} + (a+b-1)$

subject to $\lambda > 0$

(e) From Lagrangian optimality,

$$\frac{\partial \Lambda(\mathbf{y}^*, \lambda)}{\partial y_1} = -a + \lambda^* e^{y_1^*} = 0 \Rightarrow y_1^* = \ln \frac{a}{\lambda^*},$$

$$\frac{\partial \Lambda(\mathbf{y}^*, \lambda)}{\partial y_2} = -b + \lambda^* e^{y_2^*} = 0 \Rightarrow y_2^* = \ln \frac{b}{\lambda^*},$$

From $a = \lambda^* e^{y_1^*}$ and the assumption $a > 0$, it follows that $\lambda^* = 0$ is not possible. Thus, $\lambda^* > 0$. From complimentary slackness,

$$1 = e^{y_1^*} + e^{y_2^*} = \frac{a}{\lambda^*} + \frac{b}{\lambda^*} \Rightarrow \lambda^* = a + b.$$

It follows that $(y_1^*, y_2^*) = \left[ \ln \dfrac{a}{a+b}, \ln \dfrac{b}{a+b} \right]$. In summary, the primal-dual optimal solution pair is $(y_1^*, y_2^*, \lambda^*) = \left[ \ln \dfrac{a}{a+b}, \ln \dfrac{b}{a+b}, a+b \right]$.

(f) From $y_1 = \ln x_1$ and $y_2 = \ln x_2$, the optimal solution in part (a) is $(x_1^*, x_2^*) = \left( \dfrac{a}{a+b}, \dfrac{b}{a+b} \right)$. The corresponding primal optimal cost is $f^* = -\left(\dfrac{a}{a+b}\right)^a \left(\dfrac{b}{a+b}\right)^b = -\dfrac{a^a\, b^b}{(a+b)^{a+b}}.$ □

### Solution 2.8 (Using the sensitivity information)

(a) Let $\mathbf{x} = (x_1, x_2)$ and $f(x) = -3x_1 - 4x_2$. Assume for now that the constraints $x_1 \geq 0$ and $x_2 \geq 0$ do not exist. The corresponding Lagrangian is

$$\Lambda(\mathbf{x},\, \lambda) = -3x_1 - 4x_2 + \lambda_1(2x_1 + 3x_2 - 10)$$
$$+ \lambda_2(4x_1 + 2x_2 - 10)$$

From Lagrangian optimality,

$$\frac{\Lambda(\mathbf{x}^*, \lambda^*)}{\partial x_1} = -3 + 2\lambda_1^* + 4\lambda_2^* = 0$$

$$\frac{\Lambda(\mathbf{x}^*, \lambda^*)}{\partial x_2} = -4 + 3\lambda_1^* + 2\lambda_2^* = 0$$

Solving the above simultaneous equations yields $(\lambda_1^*,\, \lambda_2^*) = (1.25,\, 0.125)$. Since $\lambda_1^*,\, \lambda_2^* > 0$, complimentary slackness yields

$$2x_1^* + 3x_2^* = 10$$
$$\Rightarrow\quad (x_1^*,\, x_2^*) = (1.25,\, 2.5).$$
$$4x_1^* + 2x_2^* = 10$$

The associated primal optimal cost is $f^* = -13.75$.

Since the obtained optimal solution satisfies the constraints $x_1 \geq 0$ and $x_2 \geq 0$, it follows that solving the problem with these constraints will yield the same optimal solution. In conclusion, the primal-dual optimal solution pair is $(x_1^*, x_2^*, \lambda_1^*, \lambda_2^*) = (1.25,\, 2.5,\, 1.25,\, 0.125)$.

(b) Since $(\lambda_1^*,\, \lambda_2^*) = (1.25,\, 0.125)$, the Lagrange multipliers suggest hiring A for the extra hour. Repeating the above process, the optimal solution and the optimal cost can be found as shown below.

$$2x_1^* + 3x_2^* = 11$$
$$\Rightarrow\quad (x_1^*,\, x_2^*) = (1,\, 3),\, f^* = -15.$$
$$4x_1^* + 2x_2^* = 10$$

Similar, when the extra hour is assigned to B, the optimal solution and the optimal cost can be found as shown below.

$$2x_1^* + 3x_2^* = 10$$
$$\Rightarrow (x_1^*, x_2^*) = (1.625, 2.25), f^* = -13.875.$$
$$4x_1^* + 2x_2^* = 11$$

The above solutions verify that it is better to hire A than to hire B for the extra hour. ☐

## Solution 2.9 (Steepest descent and Newton methods)

(a) Let $\mathbf{x} = (x_1, x_2)$. It is straightforward to compute the Hessian

$$\nabla^2 f(\mathbf{x}) = \begin{bmatrix} 12x_1^2 + 2x_2^2 & 4x_1 x_2 \\ 4x_1 x_2 & 2x_1^2 + 12x_2^2 \end{bmatrix},$$

which is positive semi-definite since

$$\det([12x_1^2 + 2x_2^2]) = 12x_1^2 + 2x_2^2 \geq 0,$$
$$\det \nabla^2 f(\mathbf{x}) = 4(6x_1^4 + 6x_2^4 + 33x_1^2 x_2^2) \geq 0.$$

From $\nabla^2 f(\mathbf{x}) \geq 0$ for all $\mathbf{x}$, it follows that $f$ is convex.

(b) Note that $\nabla f(\mathbf{x}) = \begin{bmatrix} 4x_1^3 + 2x_1 x_2^2 \\ 4x_2^3 + 2x_1^2 x_2 \end{bmatrix}$. For steepest descent with stepsize equal to 0.1, $\mathbf{x}^{k+1} = \mathbf{x}^k - 0.1 \nabla f(\mathbf{x}^k)$. It follows that

$$\mathbf{x}^1 = \begin{bmatrix} 1 \\ 1 \end{bmatrix} - 0.1 \begin{bmatrix} 6 \\ 6 \end{bmatrix} = \begin{bmatrix} 0.4 \\ 0.4 \end{bmatrix},$$

$$\mathbf{x}^2 = \begin{bmatrix} 0.4 \\ 0.4 \end{bmatrix} - 0.1 \begin{bmatrix} 0.384 \\ 0.384 \end{bmatrix} = \begin{bmatrix} 0.3616 \\ 0.3616 \end{bmatrix}.$$

(c) For the Newton method with stepsize equal to 0.1, $\mathbf{x}^{k+1} = \mathbf{x}^k - 0.1(\nabla^2 f(\mathbf{x}^k))^{-1} \nabla f(\mathbf{x}^k)$.
It follows that

$$\mathbf{x}^1 = \begin{bmatrix} 1 \\ 1 \end{bmatrix} - 0.1 \begin{bmatrix} 14 & 4 \\ 4 & 14 \end{bmatrix}^{-1} \begin{bmatrix} 6 \\ 6 \end{bmatrix} = \begin{bmatrix} 0.9667 \\ 0.9667 \end{bmatrix},$$

$$\mathbf{x}^2 = \begin{bmatrix} 0.9667 \\ 0.9667 \end{bmatrix} - 0.1 \begin{bmatrix} 13.0822 & 3.7378 \\ 3.7378 & 13.0822 \end{bmatrix}^{-1} \begin{bmatrix} 5.4198 \\ 5.4198 \end{bmatrix} = \begin{bmatrix} 0.9344 \\ 0.9344 \end{bmatrix}. \quad \square$$

**Solution 2.10 (Conditional gradient and gradient projection methods)**

(*a*) From

$$\nabla f(x_1, x_2) = \begin{bmatrix} 2x_1 - x_2 \\ -x_1 + 2x_2 \end{bmatrix} \quad \text{and} \quad \nabla^2 f(x_1, x_2) = \begin{bmatrix} 2 & -1 \\ -1 & 2 \end{bmatrix},$$

it follows that $\nabla^2 f \geq 0$. Hence, $f$ is convex.

(*b*) The next point $\mathbf{x}^1$ is computed below.

$$\mathbf{x}^1 = \mathbf{x}^0 - \alpha^0 \, \nabla f(\mathbf{x}^0) = \begin{bmatrix} 1 \\ 2 \end{bmatrix} - 0.1 \begin{bmatrix} 0 \\ 3 \end{bmatrix} = \begin{bmatrix} 1 \\ 1.7 \end{bmatrix}$$

(*c*) Under the conditional gradient method, the descent direction is obtained from an optimal solution of the following problem.

$$\text{mimize } \nabla f(\mathbf{x}^0)^{\mathrm{T}}(\mathbf{x} - \mathbf{x}^0) = [0, \ 3] \begin{bmatrix} x_1 - 1 \\ x_2 - 2 \end{bmatrix} = 3(x_2 - 2)$$

$$\text{subject to } x_1 - x_2 \leq 0$$

$$x_1 \geq -1$$

From the objective function and the feasible set shown in figure B.1, the optimal solution for the above problem is the point with the minimum value of $x_2$, which is $(-1, 1)$ by inspection. Since the stepsize is 0.1, the next point $\mathbf{x}^1$ is

$$\mathbf{x}^1 = \mathbf{x}^0 + \alpha^0 \left( \begin{bmatrix} -1 \\ -1 \end{bmatrix} - x^0 \right)$$

$$= \begin{bmatrix} 1 \\ 2 \end{bmatrix} + 0.1 \left( \begin{bmatrix} -1 \\ -1 \end{bmatrix} - \begin{bmatrix} 1 \\ 2 \end{bmatrix} \right) = \begin{bmatrix} 0.8 \\ 1.7 \end{bmatrix}.$$

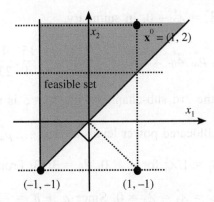

**Figure B.1** Feasible set of problem 2.10.

(*d*) Under the gradient projection method, the next point $\mathbf{x}^1$ is obtained from

$$\tilde{\mathbf{x}}^0 = (\mathbf{x}^0 - s^0 \, \nabla \, f(\mathbf{x}^0))_{|\mathcal{F}} = \left( \begin{bmatrix} 1 \\ 2 \end{bmatrix} - 1 \begin{bmatrix} 0 \\ 3 \end{bmatrix} \right)_{|\mathcal{F}} = \left( \begin{bmatrix} 1 \\ -1 \end{bmatrix} \right)_{|\mathcal{F}} = \begin{bmatrix} 0 \\ 0 \end{bmatrix},$$

where the projection is found from inspection. With the stepsize of 0.1 after the projection, the next point is

$$\mathbf{x}^1 = \mathbf{x}^0 + 0.1(\tilde{\mathbf{x}}^0 - \mathbf{x}^0)$$

$$= \begin{bmatrix} 1 \\ 2 \end{bmatrix} + 0.1 \begin{bmatrix} -1 \\ -2 \end{bmatrix} = \begin{bmatrix} 0.9 \\ 1.8 \end{bmatrix}. \qquad \square$$

**Solution 2.11 (Waterfilling)** Note that the modified noise levels (i.e. $\tilde{n}_i = n_i/h_i$) for waterfilling are equal to $(\tilde{n}_1, ..., \tilde{n}_4) = (4, 2, 5, 2)$. Let $\lambda_0^*$ be the Lagrange multiplier for the total power constraint, and $\lambda_1^*, ..., \lambda_4^*$ be the Lagrange multipliers for the inequality constraints.

(*a*) For $P = 10$, all sub-channels are allocated some power. The allocated power levels are $(p_1^*, ..., p_4^*) = \left( \dfrac{7}{14}, \dfrac{15}{4}, \dfrac{3}{4}, \dfrac{15}{4} \right)$. Since $p_i^* + \tilde{n}_i = 1/\lambda_0^*$ for $p_i^* > 0$, $\lambda_0^* = 4/23$. From complimentary slackness,

$\lambda_1^* = \lambda_2^* = \lambda_3^* = \lambda_4^* = 0$. In summary,

$$(p_1^*, \ \ldots, \ p_4^*, \ \lambda_0^*, \ \ldots, \ \lambda_4^*) = \left( \frac{7}{4}, \frac{15}{4}, \frac{3}{4}, \frac{15}{4}, \frac{4}{23}, 0, 0, 0, 0 \right).$$

(b) For $P = 5$, the 3rd sub-channel with $\bar{n}_3 = 5$ is not allocated any power. The allocated power levels are $(p_1^*, \ \ldots, \ p_4^*) = \left( \frac{1}{3}, \frac{7}{3}, 0, \frac{7}{3} \right)$. Since $p_i^* + \bar{n}_i = 1/\lambda_0^*$ for $p_i^* > 0$, $\lambda_0^* = 3/13$. From complimentary slackness, $\lambda_1^* = \lambda_2^* = \lambda_4^* = 0$. Since $p_i^* + \bar{n}_i = \dfrac{1}{\lambda_0^* - \lambda_i^*}$ for any $i$, $\lambda_3^* = 2/65$. In summary,

$$(p_1^*, \ \ldots, \ p_4^*, \ \lambda_0^*, \ \ldots, \ \lambda_4^*) = \left( \frac{1}{3}, \frac{7}{3}, 0, \frac{7}{3}, \frac{3}{13}, 0, 0, \frac{2}{65}, 0 \right). \qquad \square$$

**Solution 2.12 (Power allocation with multiple users)**

(a) No, the given problem is not convex since the feasible set contains only integer values of $x_i^j$s and is therefore not a convex set.

(b) The number of variables is

$$\underbrace{I \times J}_{x_i^j} + \underbrace{I \times J}_{p_i^j} = 2IJ.$$

The number of constraints is

$$\underbrace{1}_{\text{bound by } P} + \underbrace{I}_{\text{bound by } 1} + \underbrace{I \times J}_{\text{bound by } P x_i^j} = 1 + I + IJ.$$

(c) From the hint, note that it does not matter which user each sub-channel is assigned to as long as the same power is allocated to the selected user. Hence, one optimal solution exists when all sub-channels are assigned to user 1. In this case, the problem is reduced to waterfilling with respect to user 1 only. It follows that one optimal power allocation is given below.

$$(x_1^1, x_2^1, x_3^1, x_4^1, x_1^2, x_2^2, x_3^2, x_4^2) = (1, 1, 1, 1, 0, 0, 0, 0)$$
$$(p_1^1, p_2^1, p_3^1, p_4^1, p_1^2, p_2^2, p_3^2, p_4^2) = (2, 4, 3, 3, 0, 0, 0, 0)$$

(*d*) As suggested in the hint, the desired statement is proved by contradiction. Suppose that, in a given optimal solution, sub-channel $i$ is assigned to user $j$ while there exists user $j'$ such that $h_i^{j'} > h_i^j$. By reassigning sub-channel $i$ to user $j'$ and setting $p_i^{j'}$ to $p_i^j$, the objective function is increased by the amount equal to

$$\ln\left(1 + \frac{h_i^{j'} p_i^j}{n_i}\right) - \ln\left(1 + \frac{h_i^j p_i^j}{n_i}\right)$$

which is strictly positive since $h_i^{j'} > h_i^j$ and the objective function is strictly increasing. This contradicts the assumption that the given solution is optimal. In conclusion, each sub-channel $i$ must be assigned to the user with the maximum $h_i^j$.

(*e*) From the conclusion in part (*d*) together with the given channel gains, sub-channels 1, 2, 3, 4 are assigned to users 1, 1, 2, 2 respectively. Note that the assignments of sub-channels 2 and 4 can be arbitrary. With these assignments, waterfilling can be performed to the subcarriers using the following modified noise powers.

$$(n_1', n_2', n_3', n_4') = \left(\frac{n_1}{h_1^1}, \frac{n_2}{h_2^1}, \frac{n_3}{h_3^2}, \frac{n_4}{h_4^2}\right) = (1, 1, 1, 2).$$

The corresponding optimal solution is given below.

$$(x_1^1, x_2^1, x_3^1, x_4^1, x_1^2, x_2^2, x_3^2, x_4^2) = (1, 1, 0, 0, 0, 0, 1, 1)$$

$$(p_1^1, p_2^1, p_3^1, p_4^1, p_1^2, p_2^2, p_3^2, p_4^2) = (3.25, 3.25, 0, 0, 0, 0, 3.25, 2.25)$$

(*f*) The modified objective function is given below.

$$\text{maximize} \min_{j \in \{1, \cdots, J\}} \sum_{i=1}^{I} \ln\left(1 + \frac{h_i^j p_i^j}{n_i}\right)$$

As a simple example in which it is not optimal to assign each sub-channel $i$ to the user with the maximum gain $h_i^j$, consider the following noise and channel gain values.

$$(n_1, n_2, n_3, n_4) = (1, 1, 1, 1)$$

$$(h_1^1, h_2^1, h_3^1, h_4^1, h_1^2, h_2^2, h_3^2, h_4^2) = (2, 2, 2, 2, 1, 1, 1, 1)$$

In this scenario, all sub-channels are assigned to user 1, yielding the data rate of zero for user 2. Under the new objective, this result is worse than assigning any single sub-channel to user 2.   □

**Solution 2.13 (Minimum delay routing)**   Let $D(x_1) = \dfrac{x_1}{2c - x_1}$, $D(x_2) =$

$\dfrac{x_2}{2c - x_2}$, and $D(x_3) = \dfrac{x_3}{2c - x_3}$. By symmetry, consider solutions of the

form $x_1 \geq x_2 = x_3$. Using the minimum first derivative length (MFDL) condition, consider two cases.

- $x_1^* > 0$, $x_2^* = x_3^* = 0$: In this case,

$$\frac{\partial D_1(t)}{\partial x_1} \leq \frac{\partial D_2(0)}{\partial x_2} \quad \text{or equivalently}$$

$$\frac{2c}{(2c - t)^2} \leq \frac{1}{c} \Rightarrow t \leq \left(2 - \sqrt{2}\right)c.$$

- $x_1^*, x_2^*, x_3^* > 0$: In this case,

$$\frac{\partial D_1(x_1^*)}{\partial x_1} = \frac{\partial D_2(x_2^*)}{\partial x_2} \quad \text{or equivalently}$$

$$\frac{2c}{(2c - x_1^*)^2} = \frac{c}{(c - x_2^*)^2}.$$

Since $x_2^* = x_3^*$ and $x_1^* + x_2^* + x_3^* = t$, substituting $x_1^* = t - 2x_2^*$ in the above equation yields

$$\frac{2c}{(2c - t + 2x_2^*)^2} = \frac{c}{(c - x_2^*)^2} \Rightarrow x_2^* = \frac{t - (2 - \sqrt{2})c}{2 + \sqrt{2}},$$

$$x_1^* = t - 2x_2^* = t - 2\left(\frac{t - (2 - \sqrt{2})c}{2 + \sqrt{2}}\right) = \frac{t + \sqrt{2}(2 - \sqrt{2})c}{1 + \sqrt{2}}.$$

In summary, the optimal routing is given below.

$$(x_1^*, x_2^*, x_3^*) =$$

$$
\begin{cases}
(t, 0, 0) & t \le (2 - \sqrt{2})c \\[2mm]
\left( \dfrac{t + \sqrt{2}(2 - \sqrt{2})c}{1 + \sqrt{2}}, \dfrac{t - (2 - \sqrt{2})c}{2 + \sqrt{2}}, \dfrac{t - (2 - \sqrt{2})c}{2 + \sqrt{2}} \right) & (2 - \sqrt{2})c < t < 4c
\end{cases}
$$

$\square$

## Solution 2.14 (Network utility maximization)

(a) Viewing the problem as a minimization problem, the objective function is $f(x_1, x_2) = e^{-x_1} + e^{-x_2} - 2$. Its Hessian is $\nabla^2 f(x_1, x_2) =$

$$\begin{bmatrix} e^{-x_1} & 0 \\ 0 & e^{-x_2} \end{bmatrix}$$ and is positive semi-definite. Hence, $f$ is convex.

(b) The Lagrangian is

$$\Lambda(x_1, x_2, \lambda_0, \lambda_1, \lambda_2) = e^{-x_1} + e^{-x_2} - 2$$
$$+ \lambda_0(x_1 + x_2 - c) - \lambda_1 x_1 - \lambda_2 x_2.$$

From Lagrangian optimality,

$$-e^{-x_1^*} + \lambda_0^* - \lambda_1^* = 0 \quad \text{and} \quad -e^{-x_2^*} + \lambda_0^* - \lambda_2^* = 0.$$

From complimentary slackness, if $x_1^* + x_2^* < c$, then $\lambda_0^* = 0$, yielding $\lambda_1^* = -e^{-x_1^*} < 0$ which is not possible. Hence, $x_1^* + x_2^* = c$. Consider now three separate cases.

1. $x_1^* = 0$, $x_2^* = c$: From complimentary slackness, $\lambda_2^* = 0$. It follows that $\lambda_0^* = e^{-c}$ and $\lambda_1^* = e^{-c} - 1 < 0$. Hence, this case is not possible.

2. $x_1^* = c$, $x_2^* = 0$: By symmetry between $x_1^*$ and $x_2^*$, this case is not possible for the same reason as in the previous case.

3. $x_1^*, x_2^* > 0$: From complimentary slackness, $\lambda_1^* = \lambda_1^* = 0$, yielding $x_1^* = x_2^* = c/2$ and $\lambda_0^* = e^{-c/2}$.

In summary, the optimal primal solution is $(x_1^*, x_2^*) = \left( \dfrac{\ln 2}{2}, \dfrac{\ln 2}{2} \right)$,

while the optimal dual solution is $(\lambda_0^*, \lambda_1^*, \lambda_2^*) = \left( 1/\sqrt{2}, 0, 0 \right)$.

(c) The Lagrangian is

$$\Lambda(x_1, x_2, \lambda_0, \lambda_1, \lambda_2) = e^{-x_1} + e^{-2x_2} - 2$$
$$+ \lambda_0(x_1 + x_2 - c) - \lambda_1 x_1 - \lambda_2 x_2.$$

From Lagrangian optimality,

$$-e^{-x_1^*} + \lambda_0^* - \lambda_1^* = 0 \quad \text{and} \quad -2e^{-2x_2^*} + \lambda_0^* - \lambda_2^* = 0.$$

From complimentary slackness, if $x_1^* + x_2^* < c$, then $\lambda_2^* = 0$, yielding $\lambda_1^* = -e^{-x_1^*} < 0$ which is not possible. Hence, $x_1^* + x_2^* = c$. Consider now two separate cases. (Note that, with $u_2(c) > u_1(c)$, the case with $x_1^* = c$ and $x_2^* = 0$ need not be considered.)

1. $x_1^* = 0$, $x_2^* = c$: From complimentary slackness, $\lambda_2^* = 0$. It follows that $\lambda_0^* = 2e^{-2c}$ and $\lambda_1^* = 2e^{-2c} - 1 = 2e^{-2\ln 2} - 1 = -0.5$. Hence, this case is not possible.

2. $x_1^*, x_2^* > 0$: From complimentary slackness, $\lambda_1^* = \lambda_2^* = 0$. It follows that $e^{-x_1^*} = 2e^{-2(c-x_1^*)} = 2e^{-2(\ln 2 - x_1^*)}$, yielding $x_1^* = \dfrac{\ln 2}{3}$, $x_2^* = \dfrac{2\ln 2}{3}$ and $\lambda_0^* = e^{-(\ln 2)/3}$.

In summary, the optimal primal solution is $(x_1^*, x_2^*) = \left(\dfrac{\ln 2}{3}, \dfrac{2\ln 2}{3}\right)$, while the optimal dual solution is $(\lambda_0^*, \lambda_1^*, \lambda_2^*) = \left(1/\sqrt[3]{2}, 0, 0\right)$.

(d) From $e^{-x_1^*} = 2e^{-2(c-x_1^*)}$, the value of $c$ at which $x_1^* = 0$ is the desired answer, which is equal to $\dfrac{\ln 2}{2}$. $\qquad\qquad\square$

†**Solution 2.15 (Capacity of a DMC)** For notational simplicity, the ranges of summations are omitted in what follows. In addition, no asterisk is used to denote an optimal decision variable.

(a) The Lagrangian for the maximization problem is

$$\Lambda = \sum_x \sum_y f(y|x) f(x) \ln f(y|x) - \sum_x \sum_y f(y|x) f(x)$$

$$\ln \sum_{x'} f(y|x') f(x') + \mu\left(1 - \sum_{x} f(x)\right) + \sum_{x} \lambda_x f(x).$$

From Lagrangian optimality,

$$0 = \frac{\partial \Lambda}{\partial f(x)} = \sum_y f(y|x) \ln f(y|x)$$

$$-\sum_y \left( f(y|x) \ln \sum_{x'} f(y|x') f(x') + \frac{f(y|x)^2 f(x)}{\sum_{x'} f(y|x') f(x')} \right)$$

$$-\sum_y \sum_{x'' \neq x} f(y|x'') f(x'') \frac{f(y|x)}{\sum_{x'} f(y|x') f(x')}$$

$$-\mu + \lambda_x.$$

Noting that $\sum_{x'} f(y|x') f(x') = f(y)$, the condition is simplied to

$$0 = \sum_y f(y|x) \ln \frac{f(y|x)}{f(y)} - \sum_y \sum_{x''} f(y|x'') f(x'') \frac{f(y|x)}{f(y)} - \mu + \lambda_x$$

$$= \sum_y f(y|x) \ln \frac{f(y|x)}{f(y)} - 1 - \mu + \lambda_x,$$

where the second equality follows from the fact that $\sum_{x''} f(y|x'')$ $f(x'') = f(y)$ and $\sum_y f(y|x) = 1$. From complimentary slackness, if $f(x) > 0$, $\lambda_x = 0$, yielding

$$I(x) = \sum_y f(y|x) \ln \frac{f(y|x)}{f(y)} = 1 + \mu.$$

On the other hand, if $f(x) = 0$,

$$I(x) = \sum_y f(y|x) \ln \frac{f(y|x)}{f(y)} = 1 + \mu - \lambda_x \leq 1 + \mu.$$

Since the objective function is $\sum_x f(x) I(x)$, it follows that the optimal cost is $C = 1 + \mu$, yielding the desired statement.

(*b*) By symmetry, an optimal PMF has the following form.

$$f(x_1) = f(x_4) = p, \ f(x_2) = f(x_3) = \frac{1}{2} - p$$

From the given $f(y|x)$,

$$f(y_1) = f(y_4) = p + \frac{\varepsilon}{2} - 2p\varepsilon,$$

$$f(y_2) = f(y_3) = \frac{1}{2} - p - \frac{\varepsilon}{2} + 2p\varepsilon.$$

Using the KKT conditions in part (*a*), by setting $I(x_1) = I(x_2)$, an optimal value of $p$ can be solved from

$$(1 - \varepsilon) \ln \frac{1-\varepsilon}{f(y_1)} + \varepsilon \ln \frac{\varepsilon}{f(y_2)}$$

$$= \varepsilon \ln \frac{\varepsilon}{f(y_1)} + (1 - 2\varepsilon) \ln \frac{1-2\varepsilon}{f(y_2)} + \varepsilon \ln \frac{\varepsilon}{f(y_3)}$$

From $f(y_2) = f(y_3)$, the above condition is simplified to

$$g(p) = (1 - \varepsilon) \ln \frac{1-\varepsilon}{f(y_1)} - \varepsilon \ln \frac{\varepsilon}{f(y_1)} - (1 - 2\varepsilon) \ln \frac{1-2\varepsilon}{f(y_2)} = 0,$$

Figure B.2 shows the graph of $g(p)$ for $\varepsilon = 0.1$. From the graph, the DMC capacity is achieved by setting $p = 0.31$, yielding $C = I(x_1)|_{p=0.31} = 0.92$.[1]    □

**Solution 2.16 (True or false)**

(*a*) **TRUE:**  Suppose that **x** and **y** are two different feasible solutions. Since the feasible set is convex, any convex combination of **x** and **y** is feasible. Since the number of such convex combinations is infinite, it follows that there are infinitely many feasible solutions.

(*b*) **TRUE:**  Suppose that **x** and **y** are two different optimal solutions. Consider any convex combination $\mathbf{z} = \alpha\mathbf{x} + (1 - \alpha)\mathbf{y}$ with

---

[1]With the natural logarithm, the capacity is in nat per channel use, where "nat" stands for "natural unit". The answer in bit per channel use would be 1.33.

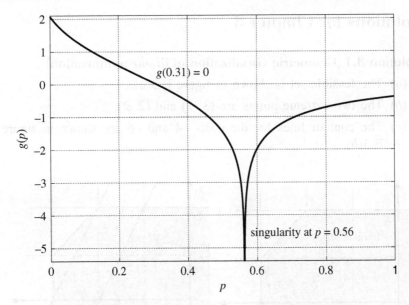

**Figure B.2** Function $g(p)$ for solving problem 2.15.

$\alpha \in [0, 1]$. Since the feasible set is convex, $\mathbf{z}$ is feasible. Let $f$ denote the objective function and $f^*$ denote the optimal cost. Since $f$ is convex,

$$f(\mathbf{z}) = f(\alpha \mathbf{x} + (1 - \alpha)\, \mathbf{y}) \leq \alpha f(\mathbf{x}) + (1 - \alpha)\, f(\mathbf{y}) = f^*,$$

which implies that $\mathbf{z}$ is also optimal. Since the number of such convex combinations is infinite, it follows that there are infinitely many optimal solutions.

(c) **FALSE:** Consider the problem of minimizing $1/x$ subject to $x \geq 1$. Although $1/x$ is bounded in the interval $[0, 1]$, there is no optimal solution.

(d) **FALSE:** Consider $f(x) = x$ and $g(x) = x^2$, both of which are convex functions over $\mathbb{R}$. The function $h(x) = f(x)\, g(x) = x^3$ is not convex over $\mathbb{R}$.

(e) **FALSE:** Consider the problem of minimizing $x_2$ subject to $x_1 \in [0, 1]$ and $x_2 \in [0, 1]$. It is easy to see that $\mathbf{x}_1 = (0, 0)$ and $\mathbf{x}_2 = (1, 0)$ are optimal. However, $\mathbf{x}_1 - \mathbf{x}_2 = (-1, 0)$ is not feasible. □

# Solutions for Chapter 3

**Solution 3.1 (Geometric visualization of linear optimization)**

(a) The feasible set is shown in figure B.3(a).

(b) The two extreme points are (3, 1) and (2, 2).

(c) The contour lines for the costs −4 and −6 are shown in figure B.3(b).

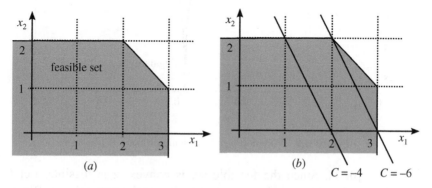

**Figure B.3**    Feasible set and contour lines of problem 3.1.

(d) By inspection, the optimal solution is (3, 1). The associated optimal cost is −7. This optimal solution is unique.

(e) Let $c_1$ and $c_2$ be cost coefficients. There are infinitely many optimal solutions for $(c_1, c_2) = (-1, -1)$.

(f) There is no optimal solution for $(c_1, c_2) = (1, 1)$.

**Solution 3.2 (Computation of BFSs, feasible directions, and reduced costs)**

(a) The transformation to the standard-form problem is given below.

$$
\begin{array}{llll}
\text{minimize} & -5x_1 - 3x_2 & \text{minimize} & -5x_1 - 3x_2 \\
\text{subject to} & 2x_1 + x_2 \le 8 & \Rightarrow \text{subject to} & 2x_1 + x_2 + s_1 = 8 \\
& x_1 + 2x_2 \le 6 & & x_1 + 2x_2 + s_2 = 6 \\
& x_1, x_2 \ge 0 & & x_1, x_2, s_1, s_2 \ge 0
\end{array}
$$

(b) The basic solution with $x_2$ and $x_1$ being basic variables is $(x_1, x_2, s_1, s_2) = (0, 3, 5, 0)$. The basic solution with $x_2$ and $s_2$ being basic variables is $(0, 8, 0, -10)$. Therefore, the only BFS is $(0, 3, 5, 0)$.

(c) The basic solution in which $x_1$ and $x_2$ are basic variables is $(x_1, x_2, s_1, s_2) = (10/3, 4/3, 0, 0)$. Since this basic solution is feasible, it is a BFS.

(d) Consider the BFS $(x_1, x_2, s_1, s_2) = (0, 3, 5, 0)$. The basic directions and reduced costs for nonbasic variables are $(1, -1/2, -3/2, 0)$ and $-7/2$ for $x_1$, and $(0, -1/2, 1/2, 1)$ and $3/2$ for $s_2$. $\square$

## Solution 3.3 (Linearization of optimization problems)

(a) Since minimizing max $(\mathbf{c}_1^T\mathbf{x}, \ldots, \mathbf{c}_K^T\mathbf{x})$ is equivalent to minimizing $z$, where $z \geq \mathbf{c}_k^T\mathbf{x}$ for all $k \in \{1, \ldots, K\}$, the problem can be transformed as follows.

| minimize max $(\mathbf{c}_1^T\mathbf{x}, \ldots, \mathbf{c}_K^T\mathbf{x})$ | | minimize $z$ |
|---|---|---|
| subject to $\mathbf{Ax} \geq \mathbf{b}$ | $\Rightarrow$ | subject to $z \geq \mathbf{c}_k^T\mathbf{x}$, $k \in \{1, \ldots, K\}$ |
| | | $\mathbf{Ax} \geq \mathbf{b}$ |

(b) Since minimizing $|\mathbf{c}^T\mathbf{x}|$ is equivalent to minimizing $z$, where $z \geq |\mathbf{c}^T\mathbf{x}|$ or equivalently $-z \leq \mathbf{c}^T\mathbf{x} \leq z$, the problem can be transformed as follows.

| minimize $|\mathbf{c}^T\mathbf{x}|$ | | minimize $z$ |
|---|---|---|
| subject to $\mathbf{Ax} \geq \mathbf{b}$ | $\Rightarrow$ | subject to $\mathbf{c}^T\mathbf{x} \leq z$ |
| | | $\mathbf{c}^T\mathbf{x} \geq -z$ |
| | | $\mathbf{Ax} \geq \mathbf{b}$ |

(c) Using the result of part (b), by introducing an extra variable $z_i$ for each term $|x_i|$, the problem can be transformed as follows.

| minimize $\sum_{i=1}^{N} c_i|x_i|$ | | minimize $\sum_{i=1}^{N} c_i z_i$ |
|---|---|---|
| subject to $\mathbf{Ax} \geq \mathbf{b}$ | $\Rightarrow$ | subject to $x_i \leq z_i$, $i \in \{1, \ldots, N\}$ |
| | | $x_i \geq -z_i$, $i \in \{1, \ldots, N\}$ |
| | | $\mathbf{Ax} \geq \mathbf{b}$ $\square$ |

**Solution 3.4 (Simplex algorithm)**

(a) The standard-form problem is given below.

$$\text{minimize} - x_1 - 2x_2$$

$$\text{subject to} - x_1 + x_2 + x_3 = 2$$

$$x_1 + x_2 + x_4 = 4$$

$$x_1, x_2, x_3, x_4 \geq 0$$

When $x_1$ and $x_2$ are nonbasic variables, the basis matrix is the identity matrix, yielding the BFS $\mathbf{x} = (0, 0, 2, 4)$.

(b) The simplex algorithm is carried out as follows.

Iteration 1

|          |   | $x_1$ | $x_2$ | $x_3$ | $x_4$ |
|----------|---|-------|-------|-------|-------|
|          | 0 | $-1$  | $-2$  | 0     | 0     |
| $x_3 =$  | 2 | $-1$  | 1     | 1     | 0     |
| $x_4 =$  | 4 | $1^*$ | 1     | 0     | 1     |

Iteration 2

|          |   | $x_1$ | $x_2$ | $x_3$ | $x_4$ |
|----------|---|-------|-------|-------|-------|
|          | 4 | 0     | $-1$  | 0     | 1     |
| $x_3 =$  | 6 | 0     | $2^*$ | 1     | 0     |
| $x_1 =$  | 4 | 1     | 1     | 0     | 1     |

Iteration 3

|          |   | $x_1$ | $x_2$ | $x_3$ | $x_4$ |
|----------|---|-------|-------|-------|-------|
|          | 7 | 0     | 0     | 0.5   | 1     |
| $x_2 =$  | 3 | 0     | 1     | 0.5   | 0     |
| $x_1 =$  | 1 | 1     | 0     | $-0.5$| 1     |

At this point, $\mathbf{x}^* = (1, 3, 0, 0)$ is obtained as an optimal solution. The associated optimal cost is $f^* = -7$.

(c) The feasible set of the problem and the path taken by the simplex algorithm are shown in figure B.4.                                        □

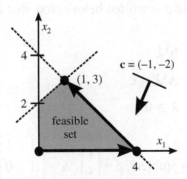

**Figure B.4** Feasible set and solution path taken by the simplex algorithm in problem 3.4.

## Solution 3.5 (Two-phase simplex algorithm)

(*a*) The starting table for phase 1 is shown below. Note that the constraints have been modified so that $\mathbf{b} \geq \mathbf{0}$.

| Iteration 1 | | $x_1$ | $x_2$ | $x_3$ | $y_1$ | $y_2$ |
|---|---|---|---|---|---|---|
| | $-6$ | 2 | 2 | 1 | 0 | 0 |
| $y_1 =$ | 1 | $-1$ | $-1$ | $-1$ | 1 | 0 |
| $y_2 =$ | 5 | $-1$ | $-1$ | 0 | 0 | 1 |

The above table already yields an optimal solution for phase 1. Since the optimal cost for phase 1 is nonzero, it follows that the problem is infeasible.

(*b*) Although the problem is infeasible, there is a basic solution since a basis can be found. For example, by choosing $x_1$ and $x_3$ as basic variables, the corresponding basic solution is $(x_1, x_2, x_3) = (-5, 0, 4)$. □

## Solution 3.6 (Duality of linear optimization)

(*a*) The Lagrangian is written as

$$\Lambda(\mathbf{x}, \lambda) = \mathbf{c}^T\mathbf{x} + \lambda^T(\mathbf{b} - \mathbf{A}\mathbf{x}).$$

(*b*) Assuming that $\mathbf{x} \in \mathbb{R}^N$, the dual function is computed below.

$$q(\lambda) = \inf_{\mathbf{x} \in \mathbb{R}^N} \Lambda(\mathbf{x}, \lambda) = \inf_{\mathbf{x} \in \mathbb{R}^N} (\mathbf{c} - \mathbf{A}^T\lambda)^T\mathbf{x} + \mathbf{b}^T\lambda$$

$$= \begin{cases} \mathbf{b}^T\lambda, & \mathbf{A}^T\lambda = \mathbf{c} \\ -\infty, & \mathbf{A}^T\lambda \neq \mathbf{c} \end{cases}$$

(c) The dual problem is written below. Note that it is a standard-form problem.

maximize $\mathbf{b}^T \boldsymbol{\lambda}$

subject to $\mathbf{A}^T \boldsymbol{\lambda} = \mathbf{c}$

$\boldsymbol{\lambda} \geq \mathbf{0}$

(d) From the given problem, $\mathbf{c} = \begin{bmatrix} 1 \\ 2 \end{bmatrix}$, $\mathbf{A} = \begin{bmatrix} 1 & 1 \\ 1 & 0 \\ -1 & 1 \end{bmatrix}$, and $\mathbf{b} = \begin{bmatrix} 1 \\ -1 \\ -4 \end{bmatrix}$. The corresponding dual problem is given below. Note that there are three dual variables, i.e. $\boldsymbol{\lambda} = (\lambda_1, \lambda_2, \lambda_3)$.

maximize $\lambda_1 - \lambda_2 - 4\lambda_3$

subject to $\lambda_1 + \lambda_2 - \lambda_3 = 1$

$\lambda_1 + \lambda_3 = 2$

$\lambda_1, \lambda_2, \lambda_3 \geq 0$    □

**Solution 3.7 (Duality of dual problem)**

(a) The dual problem is as follows.

maximize $b_1\mu_1 + b_2\mu_2$

subject to $a_{11}\mu_1 + a_{21}\mu_2 + \lambda_1 = c_1$

$a_{12}\mu_1 + a_{22}\mu_2 + \lambda_2 = c_2$

$\lambda_1, \lambda_2 \geq 0$

(b) The dual problem in the standard form is as follows.

minimize $-b_1\mu_1^+ + b_1\mu_1^- - b_2\mu_2^+ + b_2\mu_2^-$

subject to $a_{11}\mu_1^+ - a_{11}\mu_1^- + a_{21}\mu_2^+ - a_{21}\mu_2^- + \lambda_1 = c_1$

$a_{12}\mu_1^+ - a_{12}\mu_1^- + a_{22}\mu_2^+ - a_{22}\mu_2^- + \lambda_2 = c_2$

$\mu_1^+, \mu_1^-, \mu_2^+, \mu_2^-, \lambda_1, \lambda_2 \geq 0$

(c) The dual of the dual problem in part (b) is as follows.

$$\text{maximize } c_1 y_1 + c_2 y_2$$
$$\text{subject to } a_{11} y_1 + a_{12} y_2 \leq -b_1$$
$$- a_{11} y_1 - a_{12} y_2 \leq b_1$$
$$a_{21} y_1 + a_{22} y_2 \leq -b_2$$
$$- a_{21} y_1 - a_{22} y_2 \leq b_2$$
$$y_1, y_2 \leq 0$$

$\Rightarrow$

$$\text{maximize } c_1 y_1 + c_2 y_2$$
$$\text{subject to } - a_{11} y_1 - a_{12} y_2 = b_1$$
$$- a_{21} y_1 - a_{22} y_2 = b^2$$
$$y_1, y_2 \leq 0$$

(d) Using a change of variables with $z_1 = -y_1$ and $z_2 = -y_2$, the problem in part (c) can be written as

$$\text{minimize } c_1 z_1 + c_2 z_2$$
$$\text{subject to } a_{11} z_1 + a_{12} z_2 = b_1$$
$$a_{21} z_1 + a_{22} z_2 = b_2$$
$$z_1, z_2 \geq 0$$

and is the same as the given primal problem. $\qquad\square$

**Solution 3.8 (Minimum cost routing with limited budget)**

(a) Note that there are 6 s-d pairs each with 2 candidate paths. In addition, there are 14 directional links. The total number of variables is

$$\underbrace{6 \times 2}_{x^p} = 12.$$

The total number of constraints is

$$\underbrace{6}_{\forall s} + \underbrace{14}_{\forall l} + \underbrace{6 \times 2}_{\forall s, \forall p} = 32.$$

(b) The additional given parameters are listed below.

   – $B$: limited budget on the total cost of using network links

   – $\beta^s$: revenue gained for each traffic unit supported for s-d pair $s$

The reformulated optimization problem is as follows.

$$\text{maximize} \quad \sum_{s \in S} \beta^s \left( \sum_{p \in P^s} x^p \right)$$

$$\text{subject to} \quad \forall s \in S, \ \sum_{p \in P^s} x^p \le t^s$$

$$\forall l \in \mathcal{L}, \ \sum_{p \in P^s} x^p \le c_l$$

$$\sum_{l \in \mathcal{L}} \alpha_l \left( \sum_{p \in P_l} x^p \right) \le B$$

$$\forall s \in S, \ \forall_p \in P^s, \ x^p \ge 0$$

Note that the first set of constraints are modified so that the supported traffic may be less than the traffic demand. In addition, the constraint $\sum_{l \in \mathcal{L}} \alpha_l \left( \sum_{p \in P_l} x^p \right) \le B$ is added to take care of the limited budget.

(c) The total number of variables is

$$\underbrace{6 \times 2}_{x^p} = 12.$$

The total number of constraints is

$$\underbrace{6}_{\forall s} + \underbrace{14}_{\forall l} + \underbrace{1}_{\text{budget constraint}} + \underbrace{6 \times 2}_{\forall s, \forall p} = 33. \qquad \square$$

# Solutions for Chapter 4

**Solution 4.1 (Problem formulation)**   Let decision variable $x_m$, $m \in \{1, \ldots, M\}$ be the number of type-$m$ links established. The optimization problem is formulated as shown below.

$$\text{minimize} \quad \sum_{m=1}^{M} c_m x_m$$

$$\text{subject to} \quad \sum_{m=1}^{M} g_m x_m \geq R$$

$$\forall m \in \{1, \ldots, M\}, x_m \in \mathbb{Z}^+ \qquad \square$$

**Solution 4.2 (Problem formulation)**   Assume that there are $M$ days available so that the problem is always feasible. (If there are $Q < M$ days available, the problem with $M$ days can first be solved to check whether the optimal cost is greater than $Q$. If so, the problem is infeasible.) Define the following variables.

- $x_{jk} \in \{0, 1\}$: equal to 1 if and only if the exam for course $j$ is scheduled on day $k$
- $y_k \in \{0, 1\}$: equal to 1 if and only if there is at least one exam on day $k$

The following constraints force each exam to be scheduled on exactly one day.

$$\forall j \in \{1, \ldots, M\}, \sum_{k=1}^{M} x_{jk} = 1$$

The following constraints force each student to have no more than one exam on each day that is allocated for exams.

$$\forall i \in \{1, \ldots, N\}, k \in \{1, \ldots, M\}, \sum_{j=1}^{M} a_{ij} x_{jk} \leq y_k$$

The objective is to minimize the number of days with exams, i.e. minimizing $\sum_{k=1}^{M} y_k$. The overall optimization problem is as follows.

$$\text{minimize } \sum_{k=1}^{M} y_k$$

$$\text{subject to } \forall j \in \{1, ..., M\}, \sum_{k=1}^{M} x_{jk} = 1$$

$$\forall i \in \{1, ..., N\}, k \in \{1, ..., M\}, \sum_{j=1}^{M} a_{ij} x_{jk} \le y_k$$

$$\forall j, k \in \{1, ..., M\}, x_{jk} \in \{0, 1\}$$

$$\forall j \in \{1, ..., M\}, y_j \in \{0, 1\} \qquad \square$$

**Solution 4.3 (Problem formulation)**

(*a*) Define the following variables.

  - $x_{mn} \in \{0, 1\}$: equal to 1 if and only if frequency set $m$ is assigned to cell $n$
  - $y_m \in \{0, 1\}$: equal to 1 if and only if frequency set $m$ is used

The overall problem is as follows.

$$\text{minimize } \sum_{m=1}^{M} y_m$$

$$\text{subject to } \forall n \in \{1, ..., N\}, \sum_{m=1}^{M} x_{mn} = 1$$

$$\forall m \in \{1, ..., M\}, \sum_{n=1}^{N} x_{mn} \le N y_m$$

$$\forall m \in \{1, ..., M\}, \forall n, n' \in \{1, ..., N\}, a_{mn}(x_{mn} + x_{mn'}) \le 1$$

$$\forall m \in \{1, ..., M\}, \forall n \in \{1, ..., N\}, x_{mn} \in \{0, 1\}$$

$$\forall m \in \{1, ..., M\}, y_m \in \{0, 1\}$$

The first set of constraints specify that each cell is assigned exactly one frequency set. The second set of constraints specify that a frequency set can be assigned only when it is used. The third set of constraints specify that, for each pair of adjacent cells, each frequency set can be assigned at most once.

(b) The number of variables is

$$\underbrace{M \times N}_{x_{mn}} + \underbrace{M}_{y_m} = MN + M.$$

The number of constraints is

$$\underbrace{N}_{\forall n} + \underbrace{M}_{\forall m} + \underbrace{M \times N \times N}_{\forall m, \forall n, n'} = M + N + MN^2. \qquad \square$$

**Solution 4.4 (Branch-and-bound and Lagrangian relaxation):**

(a) The feasible set is illustrated in figure B.5. In particular, it contains three integer points: $(1, 0)$, $(0, 1)$, and $(1, 1)$. By inspection, the optimal solution is $x^* = (1, 1)$, while the optimal cost is $f^* = -3$.

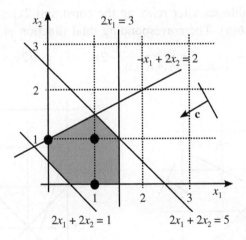

**Figure B.5** Feasible set of problem 4.4.

(b) The feasible set of the relaxed problem is the shaded region in figure B.5. By inspection, the optimal solution is $x_{relax} = (1.5, 1)$, while the optimal cost is $f_{relax} = -4$.

(c) By inspection of the relaxed problem, iteration 1 finds the optimal solution $(x_1, x_2) = (1.5, 1)$ together with additional constraints $x_1 \geq 2$ and $x_1 \leq 1$, yielding sub-problems 1 and 2.

Since the additional constraint $x_1 \geq 2$ makes the problem infeasible, sub-problem 1 need not be considered further.

For sub-problem 2 with the additional constraint $x_1 \leq 1$, the optimal solution from relaxation is $(x_1, x_2) = (1, 1.5)$ together with additional constraints $x_2 \geq 2$ and $x_2 \leq 1$, yielding sub-problems 3 and 4.

Since the additional constraint $x_2 \geq 2$ makes the problem infeasible, sub-problem 3 need not be considered further.

For sub-problem 4 with the additional constraint $x_2 \leq 1$, the optimal solution from relaxation is $(x_1, x_2) = (1, 1)$. Since this optimal solution is an integer point, $f_{best}$ and $\mathbf{x}_{best}$ are updated to $-3$ and $(1, 1)$ respectively.

Since there is no unsolved sub-problem left, in conclusion, the optimal solution is $\mathbf{x}^* = \mathbf{x}_{best} = (1, 1)$ with the optimal cost $f^* = f_{best} = -3$.

(d) The feasible set after relaxing the constraint $2x_1 \leq 3$ is shown in figure B.6(a). The corresponding dual function is

$$q(\lambda) = \min_{\mathbf{x} \in \{(1,0),(2,0),(0,1),(1,1)\}} -2x_1 - x_2 + \lambda(2x_1 - 3).$$

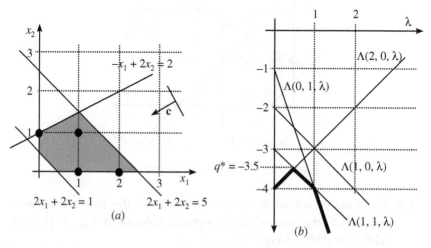

**Figure B.6**   Feasible set and dual function for Lagrangian relaxation in problem 4.4.

Figure B.6(*b*) shows the plot of the dual function. By inspection, the dual optimal cost is $q^* = -3.5$. Notice that, in this problem, $f_{\text{relax}} < q^* < f^*$. □

## Solution 4.5 (RWA for maximizing revenue):

(*a*) The missing objective is written below.

$$\text{maximize } \sum_{s \in S} r^s \left( \sum_{p \in \mathcal{P}^s} \sum_{w \in W} x_w^p \right)$$

(*b*) The number of variables of the form $x_w^p$ is

$$|\mathcal{P}| \times |\mathcal{W}| = |\mathcal{P}| \, |\mathcal{W}|.$$

The number of constraints is

$$\underbrace{|\mathcal{L} \times |\mathcal{W}|}_{\text{no collision}} + \underbrace{|S|}_{\text{traffic bound}} = |\mathcal{L}| \, |\mathcal{W}| + |S|.$$

(*c*) Yes, since there are in total $|\mathcal{P}| \, |\mathcal{W}|$ binary integer variables, there are at most $2^{|\mathcal{P}||\mathcal{W}|}$ feasible solutions. It follows that at least one of them must be optimal.

(*d*) The modified objective function is written below.

$$\text{maximize } \underbrace{\sum_{s \in S} r^s \left( \sum_{p \in \mathcal{P}^s} \sum_{w \in W} x_w^p \right)}_{\text{revenue}} - \underbrace{\sum_{l \in \mathcal{L}} \alpha_l \left( \sum_{w \in W} \sum_{p \in \mathcal{P}_l} x_w^p \right)}_{\text{cost}}$$

(*e*) Yes, **x** is always feasible since it corresponds to a subset of traffic flows with respect to $\hat{\mathbf{x}}$ and to $\tilde{\mathbf{x}}$ both of which contain a set of feasible traffic flows. □

## Solution 4.6 (Solving a Sudoku problem)

(*a*) The three sets of missing constraints are as follows.

2. Each digit $i \in \{1, \ldots, 9\}$ appears exactly once in each row.

$$\forall i, j \in \{1, \ldots, 9\}, \ \sum_{k=1}^{9} x_{ijk} = 1$$

3. Each digit $i \in \{1, ..., 9\}$ appears exactly once in each column.

$$\forall i, k \in \{1, ..., 9\}, \ \sum_{j=1}^{9} x_{ijk} = 1$$

4. Some table entries are already specified.

column 1 : $x_{411} = 1$, $x_{631} = 1$
column 2 : $x_{942} = 1$, $x_{552} = 1$, $x_{362} = 1$
column 3 : $x_{373} = 1$, $x_{283} = 1$, $x_{893} = 1$
column 4 : $x_{774} = 1$, $x_{184} = 1$, $x_{294} = 1$
column 5 : $x_{645} = 1$, $x_{355} = 1$, $x_{465} = 1$
column 6 : $x_{516} = 1$, $x_{926} = 1$, $x_{836} = 1$
column 7 : $x_{717} = 1$, $x_{127} = 1$, $x_{337} = 1$
column 8 : $x_{248} = 1$, $x_{858} = 1$, $x_{668} = 1$
column 9 : $x_{979} = 1$, $x_{599} = 1$

(b) The number of variables of the form $x_{ijk}$ is $9 \times 9 \times 9 = 729$. The number of constraints is

$$\underbrace{9 \times 9}_{\forall j, \forall k} + \underbrace{9 \times 9}_{\forall i, \forall j} + \underbrace{9 \times 9}_{\forall i, \forall k} + \underbrace{9 \times 3 \times 3}_{\forall i, \forall m, n} + \underbrace{25}_{\text{given entries}} = 349.$$

(c) If $x_{ijk}$'s and $y_{ijk}$'s are exactly the same, then $\sum_{i=1}^{9} \sum_{j=1}^{9} \sum_{k=1}^{9} y_{ijk} x_{ijk} = 81$ due to the same set of 81 variables that are equal to 1 (with all the other variables are equal to 0). Therefore, to avoid $y_{ijk}$'s to come up again as a solution, the following simple constraint can be added.

$$\sum_{i=1}^{9} \sum_{j=1}^{9} \sum_{k=1}^{9} y_{ijk} x_{ijk} \le 80 \qquad \square$$

**Solution 4.7 (True or false):**

(a) **FALSE:** Consider minimizing $x$ subject to $x \ge 0.1$, $x \le 0.9$, and $x \in \mathbb{Z}$. The relaxed problem has a feasible solution, e.g. $x = 0.5$. However, there is no feasible integer point.

(b) **TRUE:** Since the feasible set of the relaxed problem contains the feasible set of the original problem, infeasibility (i.e. empty feasible set) of the relaxed problem implies infeasibility of the original problem.

(c) **TRUE:** Let $\mathbf{x}_{int}$ be an integer feasible solution to the relaxed problem. Since $\mathbf{x}_{int}$ is also feasible in the original problem, the original problem is feasible. Let $f_{relax}$ be the optimal cost of the relaxed problem. Since $f_{relax}$ serves as a finite lower bound to the optimal cost $f^*$ for the original problem, $f^*$ is finite. Since for a linear cost function an optimal solution must exist unless the optimal cost is unbounded, it follows that there is an optimal solution.

(d) **FALSE:** Consider the counter-example shown in figure B.7. In particular, even though relaxation yields an optimal integer point (2, 2), there is a noninteger BFS (0, 2.5) in the relaxed problem.

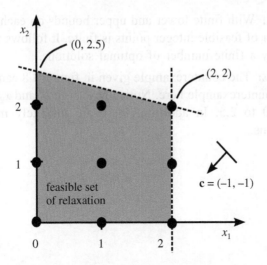

**Figure B.7** Counter-example for problem 4.7(d).

(e) **FALSE:** Consider the counter-example shown in figure B.8. In particular, even though (2, 2) is an integer optimal solution, the simplex algorithm will always return a BFS that corresponds to a corner point of the feasible set.

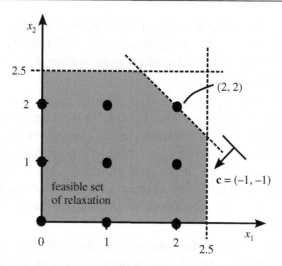

**Figure B.8**  Counter-example for problem 4.7 (*e*).

(*f*) **TRUE:** With finite lower and upper bounds on each variable, the number of feasible integer points is finite. It follows that there can be only a finite number of optimal solutions.

(*g*) **FALSE:** The counterexample given in figure B.8 can also be used as a counterexample here. Notice how both $x_1$ and $x_2$ are bounded from 0 to 2.5. In addition, there are infinitely many optimal solutions. □

# C

# Octave Commands for Optimization

Octave is a free software that can solve optimization problems [Octave, online]. In particular, the command glpk in Octave can be used to solve linear optimization problems as well as integer linear optimization problems. The instructions for using the command glpk can be obtained simply by typing help glpk at the Octave command prompt. This appendix lists specific commands related to glpk that are used to obtain numerical results in various application examples.

## C.1 Minimum Cost Routing

In example 3.9, the following octave commands are used to obtain numerical results.

```
%variable order:x12,x132,x13,x123,x14,x134,x214,x234
c    = [1;2;1;2;1;2;2;2];
A    = [1 1 0 0 0 0 0 0    ;...       %traffic demand constraint,s=1-2
        0 0 1 1 0 0 0 0    ;...       %s=1-3
        0 0 0 0 1 1 0 0    ;...       %s=1-4
        0 0 0 0 0 0 1 1    ;...       %s=2-4
        1 0 0 1 0 0 0 0    ;...       %link capacity constraint,l=(1,2)
        0 0 0 0 0 0 1 0    ;...       %l=(2,1)
        0 0 0 1 0 0 0 1    ;...       %l=(2,3)
        0 1 0 0 0 0 0 0    ;...       %l=(3,2)
        0 1 1 0 0 1 0 0    ;...       %l=(1,3)
        0 0 0 0 1 0 1 0    ;...       %l=(1,4)
        0 0 0 0 0 1 0 1]   ;          %l=(3,4)
b    = [3;4; 5;6; 5;5; 5;5; 5;5;6];
lb   = [0;0;0;0;0;0;0;0];
ub   = [];
ctype = "SSSSUUUUUUU";
[xopt fopt status extra] =glpk(c,A,b,lb,ub,ctype)
```

## C.2    Maximum Lifetime Routing in a WSN

In example 3.10, the following octave commands are used to obtain numerical results.

```
%variable order:
%x1_10,x1_20,x1_30,x1_12,x1_21,x1_13,x1_31,x1_23,x1_32
%x2_10,x2_20,x2_30,x2_12,x2_21,x2_13,x2_31,x2_23,x2_32
%x3_10,x3_20,x3_30,x3_12,x3_21,x3_13,x3_31,x3_23,x3_32
c = [0;0;0;0;0;0;0;0;0;0;  0;0;0;0;0;0;0;0;0;  0;0;0;0;0;0;0;0;0;  1];
A = [0.3 0 0 0.2 0.1 0.6 0.1 0 0 0.3 0 0 0.2 0.1 0.6 0.1 0 0...
                0.3 0 0 0.2 0.1 0.6 0.1 0 0 -1;... %lifetime constraint,i=1
     0 0.6 0 0.1 0.2 0 0 0.3 0.1 0 0.6 0 0.1 0.2 0 0 0.3 0.1...
            0 0.6 0 0.1 0.2 0 0 0.3 0.1 -1;... %i=2
     0 0 1 0 0 0.1 0.6 0.1 0.3 0 0 1 0 0 0.1 0.6 0.1 0.3...
            0 0 1 0 0 0.1 0.6 0.1 0.3 -1;... %i=3
     1 1 1 0 0 0 0 0 0 0 0 0 0 0 0 0 0 0 0 0 0 0 0 0 0 0 0 0;...
                %flow conservation constraint,i=1,k=0
     -1 0 0 -1 1 -1 1 0 0 0 0 0 0 0 0 0 0 0 0 0 0 0 0 0 0 0 0 0;... %i=1,k=1
     0 -1 0 1 -1 0 0 -1 1 0 0 0 0 0 0 0 0 0 0 0 0 0 0 0 0 0 0 0;... %i=1,k=2
     0 0 -1 0 0 1 -1 1 -1 0 0 0 0 0 0 0 0 0 0 0 0 0 0 0 0 0 0 0;... %i=1,k=3
     0 0 0 0 0 0 0 0 0 1 1 1 0 0 0 0 0 0 0 0 0 0 0 0 0 0 0 0;... %i=2,k=0
     0 0 0 0 0 0 0 0 0 -1 0 0 -1 1 -1 1 0 0 0 0 0 0 0 0 0 0 0 0;... %i=2,k=1
     0 0 0 0 0 0 0 0 0 0 -1 0 1 -1 0 0 -1 1 0 0 0 0 0 0 0 0 0 0;... %i=2,k=2
     0 0 0 0 0 0 0 0 0 0 -1 0 0 1 -1 1 -1 0 0 0 0 0 0 0 0 0 0 0;... %i=2,k=3
     0 0 0 0 0 0 0 0 0 0 0 0 0 0 0 0 0 0 1 1 1 0 0 0 0 0 0 0;... %i=3,k=0
     0 0 0 0 0 0 0 0 0 0 0 0 0 0 0 0 0 0 -1 0 0 -1 1 -1 1 0 0 0;... %i=3,k=1
     0 0 0 0 0 0 0 0 0 0 0 0 0 0 0 0 0 0 -1 0 1 -1 0 0 -1 1 0;... %i=3,k=2
     0 0 0 0 0 0 0 0 0 0 0 0 0 0 0 0 0 0 -1 0 0 1 -1 1 -1 0]; %i=3,k=3
b = [0;0;0;  1;-1;0;0;1;0;-1;0;1;0;0;-1];
ctype = "UUUSSSSSSSSSSSSS";
lb = [0;0;0;0;0;0;0;0;0;  0;0;0;0;0;0;0;0;0;  0;0;0;0;0;0;0;0;0;  0];
ub = [];
[xopt fopt status extra] = glpk(c,A,b,lb,ub,ctype)
```

# C.3 RWA in a WDM Network

In example 4.4, the following octave commands are used to obtain numerical results.

```
%variable order:
%x_1^123,x_1^143,x_1^24,x_1^214,x_1^321,x_1^341,x_1^345,x_1^3215
%x_1^42,x_1^432,x_1^45,x_1^415,x_1^543,x_1^5213,x_1^54,x_1^514
%x_2^123,x_2^143,x_2^24,x_2^214,x_2^321,x_2^341,x_2^345,x_2^3215
%x_2^42,x_2^432,x_2^45,x_2^415,x_2^543,x_2^5213,x_2^54,x_2^514
c = [2;2;1;2;2;2;2;3;1;2;1;2;2;3;1;2; 2;2;1;2;2;2;2;3;1;2;1;2;2;3;1;2;];
A = [1 0 0 0 0 0 0 0 0 0 0 0 0 1 0 0 0 0 0 0 0 0 0 0 0 0 0 0 0 0 0 0;...
                  %wavelength collision constraint,l=(1,2),w=1
1 0 0 0 0 0 0 0 0 0 0 0 1 0 0 0 0 0 0 0 0 0 0 0 0 0 0 0 0 0 0;... %l=(2,3),w=1
0 0 0 0 0 1 1 0 0 0 0 0 0 0 0 0 0 0 0 0 0 0 0 0 0 0 0 0 0 0 0;... %l=(3,4),w=1
0 0 0 0 0 0 1 0 0 0 1 0 0 0 0 0 0 0 0 0 0 0 0 0 0 0 0 0 0 0 0;... %l=(4,5),w=1
0 0 0 0 0 0 0 0 0 0 0 1 0 1 0 0 0 0 0 0 0 0 0 0 0 0 0 0 0 0 0;... %l=(5,1),w=1
0 1 0 1 0 0 0 0 0 0 0 0 0 1 0 0 0 0 0 0 0 0 0 0 0 0 0 0 0 0 0;... %l=(1,4),w=1
0 0 1 0 0 0 0 0 0 0 0 0 0 0 0 0 0 0 0 0 0 0 0 0 0 0 0 0 0 0 0;... %l=(2,4),w=1
0 0 0 1 1 0 0 1 0 0 0 0 0 0 0 0 0 0 0 0 0 0 0 0 0 0 0 0 0 0 0;... %l=(2,1),w=1
0 0 0 0 1 0 0 1 0 1 0 0 0 0 0 0 0 0 0 0 0 0 0 0 0 0 0 0 0 0 0;... %l=(3,2),w=1
0 1 0 0 0 0 0 0 1 0 0 1 0 0 0 0 0 0 0 0 0 0 0 0 0 0 0 0 0 0 0;... %l=(4,3),w=1
0 0 0 0 0 0 0 0 0 0 0 1 0 1 0 0 0 0 0 0 0 0 0 0 0 0 0 0 0 0 0;... %l=(5,4),w=1
0 0 0 0 0 0 1 0 0 0 1 0 0 0 0 0 0 0 0 0 0 0 0 0 0 0 0 0 0 0 0;... %l=(1,5),w=1
0 0 0 0 1 0 0 0 0 0 1 0 0 0 0 0 0 0 0 0 0 0 0 0 0 0 0 0 0 0 0;... %l=(4,1),w=1
0 0 0 0 0 0 0 1 0 0 0 0 0 0 0 0 0 0 0 0 0 0 0 0 0 0 0 0 0 0 0;... %l=(4,2),w=1
0 0 0 0 0 0 0 0 0 0 0 0 1 0 0 0 0 0 0 0 0 0 0 0 1 0 0;... %l=(1,2),w=2
0 0 0 0 0 0 0 0 0 0 0 0 1 0 0 0 0 0 0 0 0 0 0 1 0 0;... %l=(2,3),w=2
0 0 0 0 0 0 0 0 0 0 0 0 0 0 0 0 0 1 1 0 0 0 0 0 0 0;... %l=(3,4),w=2
0 0 0 0 0 0 0 0 0 0 0 0 0 0 0 0 0 0 1 0 0 0 1 0 0 0 0;... %l=(4,5),w=2
0 0 0 0 0 0 0 0 0 0 0 0 0 0 0 0 0 0 0 0 0 0 1 0 1;... %l=(5,1),w=2
0 0 0 0 0 0 0 0 0 0 0 0 0 1 0 1 0 0 0 0 0 0 0 0 1;... %l=(1,4),w=2
0 0 0 0 0 0 0 0 0 0 0 0 0 1 0 0 0 0 0 0 0 0 0 0 0;... %l=(2,4),w=2
0 0 0 0 0 0 0 0 0 0 0 0 0 1 1 0 0 1 0 0 0 0 0 0 0;... %l=(2,1),w=2
0 0 0 0 0 0 0 0 0 0 0 0 0 1 0 0 1 0 1 0 0 0 0 0;... %l=(3,2),w=2
0 0 0 0 0 0 0 0 0 0 0 0 1 0 0 0 0 0 1 0 0 1 0 0 0;... %l=(4,3),w=2
0 0 0 0 0 0 0 0 0 0 0 0 0 0 0 0 0 0 0 0 0 1 0 1 0;... %l=(5,4),w=2
0 0 0 0 0 0 0 0 0 0 0 0 0 0 0 0 1 0 0 0 1 0 0 0;... %l=(1,5),w=2
```

```
0 0 0 0 0 0 0 0 0 0 0 0 0 0 0 0 0 0 0 0 1 0 0 0 0 1 0 0 0 0;... %l=(4,1),w=2
0 0 0 0 0 0 0 0 0 0 0 0 0 0 0 0 0 0 0 0 0 0 0 1 0 0 0 0 0 0;... %l=(4,2),w=2
1 1 0 0 0 0 0 0 0 0 0 0 0 0 0 1 1 0 0 0 0 0 0 0 0 0 0 0 0 0;...
                 %traffic demand constraint,s=1-3
    0 0 1 1 0 0 0 0 0 0 0 0 0 0 0 0 1 1 0 0 0 0 0 0 0 0 0 0;... %s=2-4
    0 0 0 0 1 1 0 0 0 0 0 0 0 0 0 0 0 0 1 1 0 0 0 0 0 0 0 0;... %s=3-1
    0 0 0 0 0 0 1 1 0 0 0 0 0 0 0 0 0 0 0 0 1 1 0 0 0 0 0 0;... %s=3-5
    0 0 0 0 0 0 0 0 1 1 0 0 0 0 0 0 0 0 0 0 0 0 1 1 0 0 0 0;... %s=4-2
    0 0 0 0 0 0 0 0 0 0 1 1 0 0 0 0 0 0 0 0 0 0 0 0 1 1 0 0 0;... %s=4-5
    0 0 0 0 0 0 0 0 0 0 0 0 1 1 0 0 0 0 0 0 0 0 0 0 0 0 1 1 0 0;... %s=5-3
    0 0 0 0 0 0 0 0 0 0 0 0 0 0 1 1 0 0 0 0 0 0 0 0 0 0 0 0 1 1]; %s=5-4
b = [1;1;1;1;1;1;1;1;1;1;1;1;1;1;1;1;1;1;1;1;1;1;1;1;  1;1;1;1;1;1;1;1];
lb = [0;0;0;0;0;0;0;0;0;0;0;0;0;0;0;0;0;0;0;0;0;0;0;0;0;0;0;0;0;0];
ub = [1;1;1;1;1;1;1;1;1;1;1;1;1;1;1;1;1;1;1;1;1;1;1;1;1;1;1;1;1;1];
ctype = ["UUUUUUUUUUUUUUUUUUUUUUUUUSSSSSSSS"];
vartype = ["IIIIIIIIIIIIIIIIIIIIIIIIIIIIIIIII"];
[xopt fopt status extra] = glpk(c,A,b,lb,ub,ctype,vartype)
```

## C.4   Network Topology Design

In example 4.5, the following octave commands are used to obtain numerical results.

```
%variable order:  x11,x12,x13,x14,x15,x21,...,x55,y1,y2,y3,y4,y5
c = [0 sqrt(10) sqrt(5) sqrt(13) 4 sqrt(10) 0 sqrt(5) sqrt(5) sqrt(18)...
     sqrt(5) sqrt(5) 0 sqrt(2) sqrt(5) sqrt(13) sqrt(5) sqrt(2) 0 sqrt(5)...
     4 sqrt(18) sqrt(5) sqrt(5) 0 10 10 10 10 10];
A = [1 1 1 1 1 0 0 0 0 0 0 0 0 0 0 0 0 0 0 0 0 0 0 0 0 0 0 0 0 0;...
             %assignment constaint,i=1
     0 0 0 0 0 1 1 1 1 1 0 0 0 0 0 0 0 0 0 0 0 0 0 0 0 0 0 0 0 0;... %i=2
     0 0 0 0 0 0 0 0 0 0 1 1 1 1 1 0 0 0 0 0 0 0 0 0 0 0 0 0 0 0;... %i=3
     0 0 0 0 0 0 0 0 0 0 0 0 0 0 0 1 1 1 1 1 0 0 0 0 0 0 0 0 0 0;... %i=4
     0 0 0 0 0 0 0 0 0 0 0 0 0 0 0 0 0 0 0 0 1 1 1 1 1 0 0 0 0 0]; %i=5
     1 0 0 0 0 1 0 0 0 0 1 0 0 0 0 1 0 0 0 0 1 0 0 0 0 -3 0 0 0 0;...
             %capacity constraint,j=1
     0 1 0 0 0 0 1 0 0 0 0 1 0 0 0 0 1 0 0 0 0 1 0 0 0 0 -3 0 0 0;... %j=2
     0 0 1 0 0 0 0 1 0 0 0 0 1 0 0 0 0 1 0 0 0 0 1 0 0 0 0 -3 0 0;... %j=3
     0 0 0 1 0 0 0 0 1 0 0 0 0 1 0 0 0 0 1 0 0 0 0 1 0 0 0 0 -3 0; ...%j=4
     0 0 0 0 1 0 0 0 0 1 0 0 0 0 1 0 0 0 0 1 0 0 0 0 1 0 0 0 0-3]; %j=5
b = [1;1;1;1;1;0;0;0;0;0];
ctype = "SSSSSUUUUU";
lb = [0;0;0;0;0;0;0;0;0;0;0;0;0;0;0;0;0;0;0;0;0;0;0;0;0;0;0;0;0;0];
ub = [1;1;1;1;1;1;1;1;1;1;1;1;1;1;1;1;1;1;1;1;1;1;1;1;1;1;1;1;1;1];
ctype = ["UUUUUUUUUUUUUUUUUUUUUUUUUSSSSSSSSS"];
vartype = ["IIIIIIIIIIIIIIIIIIIIIIIIIIIIIIII"];
[xopt fopt status extra] = glpk(c,A,b,lb,ub,ctype,vartype)
```

# Bibliography

Apostol,T.M. (1969). *Calculus*, volume 2. John Wiley & Sons, New York, NY, USA.

Bertsekas, D. (1995). *Nonlinear Programming*. Athena Scientific, Nashua, NH, USA.

Bertsimas, D. and R.G. Gallager. (1992). *Data Networks*. Prentice Hall, Upper Saddle River, NJ, USA.

Bertsimas, D. and J.N. Tsitsikilis. (1997). *Introduction to Linear Optimization*. Athena Scientific, Nashua, NH, USA.

Boyd, S. and L. Vandenberghe. (2004). *Convex Optimization*. Cambridge University Press, Cambridge, UK.

Chang, J.H. and L. Tassiulas. (2004). Maximum lifetime routing in wireless sensor networks. *IEEE Transactions on Communications* 12(4): 609-619.

Gallager, R.G. (2008). *Principles of Digital Communication*. Cambridge University Press, Cambridge, UK.

Garey, M.R. and D.S. Johnson. (1979). *Computers and Intractability:A Guide to the Theory of NP-Completeness*. W.H. Freeman, New York, NY, USA.

Heinzelman, W., A. Chandrakasan and H. Balakrishnan. (2002). An application-specific protocol architecture for wireless microsensor networks. *IEEE Transactions on Wireless Communications* 1(4): 660-670.

Kennedy, J. and R.C. Eberhart. (1995). Particle swarm optimization. *Proceedings of IEEE International Conference on Neural Networks*. IEEE Press, Piscataway, NJ, USA.

Octave (online). www.octave.org

Papadimitriou, C.H. and K. Steiglitz. (1998). *Combinatorial Optimization*: *Algorithms and Complexity*. Dover, Mineola, NY, USA.

Phyo, M.M. (2006). Energy-Efficient Hierarchical LEACH-Based Multi-Hop Routing for Wireless Sensor Networks. MasterThesis, Asian Institute of Technology, Pathumthani, Thailand.

Proakis, J.G. and M. Salehi. (2008). *Digital Communications*, 5th Edition. McGraw-Hill, New York, NY, USA.

Robertazzi, T.G. (1999). *Planning Telecommunications Networks*. IEEE Press, Piscataway, NJ, USA.

Rudin, W. (1976). *Principles of Mathematical Analysis*. McGraw-Hill, New York, NY, USA.

Simmons, J.M. (2008). *Optical Network Design and Planning*. Springer-Verlag, New York, NY, USA.

Strang, G. (2005). *Linear Algebra and Its Applications*, 4th Edition. Brooks/Cole Publishing Co., Florence, KY, USA.

# Index

For inquiries about Datasets and Information please contact our
EU representative GPSR, Rem, srl, gpsr@remesrl.com, via G.A. Frumento
Verlag GmbH, Robert-Bosch-Ring 21, 85410 Haar near München, Germany